THE UNIVERSE...AND BEYOND

THE
UNIVERSE
...AND BEYOND

REVISED AND EXPANDED

TERENCE DICKINSON   CAMDEN HOUSE

My first thanks must go to the hundreds of people who asked questions in my astronomy classes at St. Lawrence College, in Kingston, Ontario, over the past decade and a half, giving me a good gauge of what to put in—and what to leave out. Thanks also go to Roy Bishop (Acadia University), David Hanes (Queen's University), Alan Dyer (*Astronomy* magazine), Paul Deans (MacMillan Planetarium), Paul Feldman (National Research Council) and Marshall McCall and Robert Garrison (both University of Toronto), who read and commented on all, or part, of the manuscript for the first edition, and to Barry Estabrook, my editor, for many helpful suggestions. More thanks go to designers Ulrike Bender (first edition) and Linda Menyes (this edition) and copy editors Mary Patton, Charlotte DuChene, Christine Kulyk and Catherine DeLury. Photographs and additional art and illustrations were supplied through the generosity of many individuals and institutions listed on page 153. In this regard, Ray Villard of the Space Telescope Science Institute was especially helpful. The essay on pages 116 to 119 is adapted from an article that first appeared in the *Griffith Observer*, September 1985, and is reproduced with permission. My largest debt of gratitude goes to my wife Susan, who was intimately involved in many stages of the book's production and provided invaluable advice as copy editor, proofreader and tireless aide.

© Copyright 1986, 1992 by Terence Dickinson

Second Printing 1988
Third Printing (major revision) 1992
Fourth Printing 1993

Canadian Cataloguing in Publication Data

Dickinson, Terence
  The universe — and beyond

Includes index.
ISBN 0-921820-51-8 (bound)   ISBN 0-921820-53-4 (pbk.)

1. Astronomy — Popular works.   2. Life on other planets.   I. Title.

QB44.2.D53 1992   520   C92-094743-3

Front Cover: View from a hypothetical planet toward a starburst region near the nucleus of the Milky Way Galaxy (see page 140). Illustration by Adolf Schaller. Back Cover: Geysers on Enceladus, a satellite of Saturn (see page 58). Illustration by Adolf Schaller. Frontispiece: Lagoon Nebula, a 50-light-year-wide cloud of dust and gas, 4,500 light-years from Earth. National Optical Astronomy Observatories, Kitt Peak National Observatory.

Dedication:
To Reg, Anna, Stephen and Dianne, for a lifetime of love and support.

Designed by
Ulrike Bender

Printed and bound in Canada by
D.W. Friesen & Sons Ltd.
Altona, Manitoba

Printed on acid-free paper

Published by Camden House Publishing
(a division of Telemedia Communications Inc.)

Camden House Publishing
7 Queen Victoria Road
Camden East, Ontario K0K 1J0

Camden House Publishing
Box 766
Buffalo, New York 14240-0766

Printed and distributed under exclusive licence from Telemedia Communications Inc. by
Firefly Books
250 Sparks Avenue
Willowdale, Ontario M2H 2S4

Firefly Books (U.S.) Inc.
P.O. Box 1338
Ellicott Station
Buffalo, New York 14205

CONTENTS

ASTRONOMY: EDIFICATION OR ESCAPISM?

Not long ago, I was the guest "expert" on a radio call-in show on CFRB, in Toronto. Andy Barrie, the genial host, had asked me to explain how we can see the universe's past by looking at remote galaxies whose light has taken billions of years to reach us. Barrie was well aware that people are endlessly fascinated with this topic, along with the latest ideas about the origin of the universe, which we discussed next.

We got the usual mix of calls that we both were used to, but then a caller asked: "Why do we need to know any of this stuff? What good is it? It doesn't affect our lives one bit."

One has to think fast on live radio yet, at the same time, be careful not to say something that might be regretted later. But then I realized the caller had a point, and I agreed. "You are quite right, we don't *need* to know any of it," I said, then countered with, "but haven't you ever wondered how it all happened—where Earth and the universe came from?"

The answer was firm: "Anyone can get along perfectly well in life without knowing a thing about the stars."

I don't remember the exact conversation from there, except that I tried to describe how I thought my life had been enriched by filling my mind with astronomical "stuff" that nobody really needs to know, how it buffers me from some of the unpleasant realities on this planet.

Is astronomy escapism? If so, there are lots of armchair astronomers waiting for their next fix. In any case, the same questions could be asked about music, art or classic fiction. None of it is necessary. But all of it is food for the mind, providing insight and understanding on human and cosmic scales. For those interested in the cosmic side of the equation, I invite you to soar with me among alien worlds and distant galaxies on the following pages.

Since it is always possible to improve a book, I am pleased to have the opportunity to do so in this revised and expanded edition. Changes throughout the book incorporate many significant developments in astronomy from the past half decade:

☐ The spectacular radar mapping of Venus by the Magellan spacecraft during its multi-year orbital reconnaissance of the planet.

☐ Up-to-date results from the Hubble Space Telescope, with new images of planets, nebulas and galaxies.

☐ Details of the 1989 Voyager 2 encounter with Neptune, the final planetary flyby of the century.

☐ A cosmology update section describing the major astronomical controversy that has erupted in the 1990s about the origin and evolution of the universe.

☐ An essay on the significance of astronomical art.

☐ More than 20 new photographs and illustrations.

☐ Updated facts, figures and tables throughout.

Overall, the book retains its basic form, but certain rearrangements of material and modifications in page layout have increased the actual contents by at least 10 pages. I was guided in this revision partly by comments from readers of the earlier printings. You are welcome to write to me at the Camden House address on the copyright page. I will respond to every letter.

—Terence Dickinson
Yarker, Ontario
June 1992

Every generation thinks it has the answers, and every generation is humbled by nature.

PHILIP LUBIN
Cosmologist
1992

A family portrait of our solar system includes eight planets plus the Earth's moon. Pluto, the ninth planet, is the only one yet to be seen in detail by spacecraft. As recently as 1970, none of these images existed except the Earth-based photographs of the moon and Mars. Interplanetary probes have revolutionized our knowledge of these neighbour worlds, a saga that is chronicled in the first half of this book. In the second half, more distant realms are explored, represented in these introductory pages by the Hercules Cluster, page 6, and the S Monocerotis Nebula, page 5.

CHAPTER ONE

JOURNEY THROUGH TIME AND SPACE

T he spring evening is crisp and cool and pitch-black. Stars fill the sky in a glittering tapestry that goes unnoticed by the occupants of a car speeding down the rural highway, far from city lights and traffic. Dimly at first, the headlights reveal a steep hill ahead. Without losing speed, the automobile hurtles upward and reaches the crest, and for an instant, the headlamps shine into the blackness like two ghostly fingers. That instant marks freedom for at least one photon of light which avoids bumping into dust motes or absorbing molecules of air on its way up from the Earth's surface. Less than two seconds later, it passes the moon. One minute after that, Earth and moon diminish to starlike points. Within an hour, they fade into the starry backdrop. One month away from Earth, the photon is so remote that all the planets are invisible and the sun dwindles to a star, though still far brighter than any other. In two years, the sun is reduced to a bright but not extraordinary star. Several other stars are similarly luminous.

Over the next 50 years, the sun slowly fades until it is dimmer than the faintest stars visible to the unaided eye. Yet the sky still appears basically as it does from Earth and has the same general proportion of bright and dim stars. But by the hundredth year of travel, a distinct thinning out of stars becomes apparent ahead. The photon is moving out of the Milky Way Galaxy.

After cruising in its arrow-straight trajectory for 2,000 years, the photon is completely outside and well above the spiral arm of the galaxy where our solar system resides. From this vantage point, only a handful of stars speckles the sky. But the view back toward the galaxy reveals an impressive panorama: the sweeping curves of the spiral arms and, beyond them, the bulging galactic nucleus. Continuing its voyage for another 22,000 years, the photon nears a mammoth swarm of stars, the Hercules Cluster, a million suns congregated in a rough sphere about 75 light-years across. This is one of approximately 150 globular clusters that orbit the Milky Way Galaxy like satellites.

The photon races onward, but the scenery becomes less inspiring with each millennium as the Milky Way fades to a mere puff in the blackness. Only one additional galaxy, Andromeda, is easily visible; other galaxies appear as mere smudges. After 10 million years, both the Milky Way and Andromeda are lost to view. Millions more years will pass before chance encounters with other galaxies break the monotony of the void. And the journey has just begun.

Experiencing the Cosmos

It is no accident that the starry night stirs in us the most profound questions of origins, destinies and the ultimate meaning of it all. Humans are as much a part of the fabric of the cosmos as a moon rock, an ice particle in the rings of Saturn or an asteroid in a galaxy a billion light-years away. We are all star-stuff, assemblages of atoms cooked in the thermonuclear fires at the hearts of stars. And before there were stars, every subatomic particle in the universe emerged from a genesis fireball trillions of times hotter than the sun.

Our cosmic roots are revealed in another subtle

A day will come when beings shall stand upon this Earth, as one stands upon a footstool, and laugh and reach out their hands amidst the stars.

H.G. WELLS
1902

The Horsehead Nebula, one of the sky's most distinctive landmarks, is a dark cloud of cosmic gas and dust silhouetted against a more distant cloud illuminated by the bright star at left. The Horsehead is about one light-year across and 1,500 light-years from Earth. (One light-year is 63,000 times the Earth-sun distance.)

11

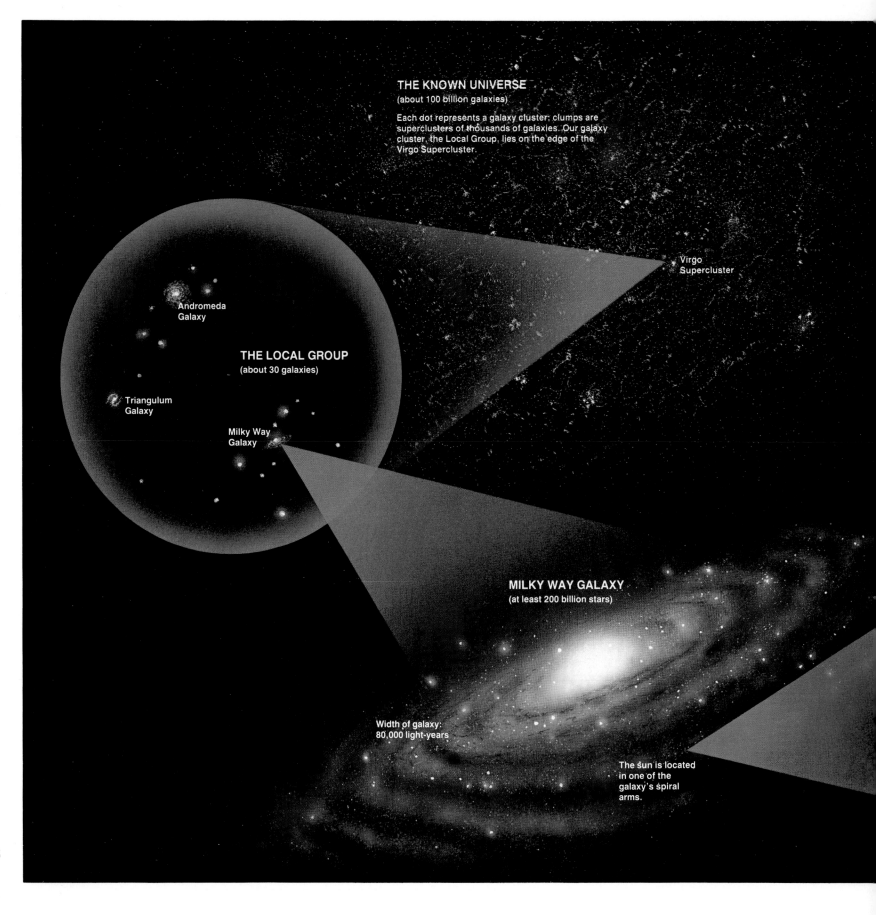

THE KNOWN UNIVERSE
(about 100 billion galaxies)

Each dot represents a galaxy cluster; clumps are superclusters of thousands of galaxies. Our galaxy cluster, the Local Group, lies on the edge of the Virgo Supercluster.

Virgo Supercluster

Andromeda Galaxy

THE LOCAL GROUP
(about 30 galaxies)

Triangulum Galaxy

Milky Way Galaxy

MILKY WAY GALAXY
(at least 200 billion stars)

Width of galaxy: 80,000 light-years

The sun is located in one of the galaxy's spiral arms.

12

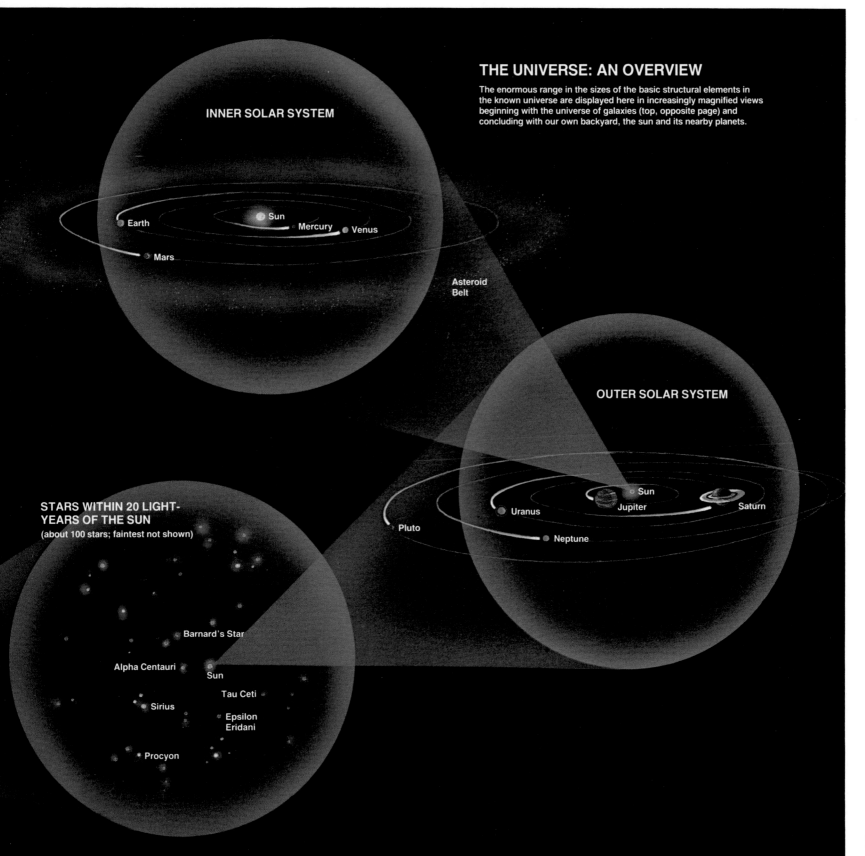

THE UNIVERSE: AN OVERVIEW

The enormous range in the sizes of the basic structural elements in the known universe are displayed here in increasingly magnified views beginning with the universe of galaxies (top, opposite page) and concluding with our own backyard, the sun and its nearby planets.

INNER SOLAR SYSTEM

Sun

Earth

Mercury

Venus

Mars

Asteroid Belt

OUTER SOLAR SYSTEM

Sun

Jupiter

Saturn

Uranus

Pluto

Neptune

STARS WITHIN 20 LIGHT-YEARS OF THE SUN

(about 100 stars; faintest not shown)

Barnard's Star

Alpha Centauri

Sun

Tau Ceti

Sirius

Epsilon Eridani

Procyon

way. Many people find that celestial objects exhibit a beauty which is difficult to qualify. Like a favourite piece of music or a painting that invites another glance, the delicate tendrils of a nebula or the pinwheeling arms of a spiral galaxy are irresistible. When Canadian astronaut Marc Garneau returned from a six-day flight on the space shuttle in 1984 and was asked what he found most memorable about the journey, he was emphatic: "The view of the Earth! It was incredibly beautiful. The hardest thing I had to do up there was tear myself away from the window." Fewer than 300 humans have seen our planet from space—a sapphire sphere dappled with cotton clouds and continents wearing vests of green and brown. Fewer still have ventured into the moon's ken, and only 12 have actually walked on its dusty, cratered face.

Beyond the moon, where no one has yet travelled, robot probes embodying electronic and computer substitutes for human senses have allowed us the vicarious exploration of more than two dozen planets and moons. The red deserts of Mars, the sulphur volcanoes of Jupiter's satellite Io, the mammoth craters of Saturn's moon Mimas and the black rings of Uranus have been revealed almost as clearly as if we had been there. More distant cosmic shores, the realm of myriad stars and galaxies, remain unexplored but not unknown. Centuries of telescopic scrutiny have provided a vast inventory of facts—distances, sizes, ages—which are buttressed by theories that attempt to decipher origins and predict destinies.

What is it like out there? What varieties of worlds wheel around those distant suns? Are there other creatures in the universe that share our compulsion to know? Those questions in their broadest context are what this book addresses. It includes frequent imaginative voyages into various cosmic environments, descriptions of what it would actually be like to be there. I have left out some of the material commonly found in astronomy textbooks or encyclopaedias, particularly discourses on instrumentation and how we learned this or that fact. There are plenty of excellent books that provide such information, and a few of the best are listed at the back of the book. The approach here is more a guided tour, a celebration of cosmic wonderment.

My primary goal has been to present the various objects and elements of the universe as they would be experienced by human explorers. We stroll the sands of Mars, float among Saturn's rings, observe how one star is born and another dies and venture to planets of double suns and realms where black holes consume nearby stars or swallow whole galaxies. We hurtle back in time to the very origin of the universe, before anything we know today existed—to the first infinitesimal fraction of a second. These expeditions are based on fact. I have endeavoured to keep our celestial excursions rooted in current scientific knowledge. Only in the discussion of extraterrestrial life in the final chapter does conjecture become a major component.

To the Edge of the Universe

The voyage begins with the vista of stars seen on a dark, moonless night. As each tiny spear of starlight enters the eyes, it ends a journey that began decades or centuries ago. Five of the seven stars in the Big Dipper, for example, are members of a nearby star cluster roughly 75 light-years away. Their light takes a human lifetime to reach Earth. Nestled between the constellations Perseus and Cassiopeia lies a dim blur of light that binoculars reveal as a twin cluster of stars—the Double Cluster. Their light travels for 7,000 years before it reaches us from the next spiral arm beyond the one that we occupy in the Milky Way Galaxy.

The unaided eye can bridge even greater gulfs. Overhead, on late-autumn evenings, we can see

the Andromeda Galaxy, a small, oval, hazy patch that is, in reality, a colossal platter-shaped stellar metropolis larger than our own Milky Way and populated by perhaps a trillion stars. The combined light of those stars is dimmed by its enormous distance from Earth: two million light-years.

When light from the Andromeda Galaxy meets a human retina and registers as an image in the brain, that particular bundle of photons terminates an uninterrupted voyage of two million years. One of the photons could have originated on the surface of a star like our sun. Light from that same star might bathe a family of planets, one of which could resemble Earth. Such thoughts are central to the lure of astronomy. The possibility of other inhabited worlds in the universe is a theme I will return to as our exploration of the cosmos unfolds.

The Andromeda Galaxy is just the nearest of billions of galaxies similar to the Milky Way. The most powerful telescopes on Earth can detect

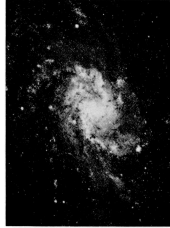

*The twirling arms of a spiral galaxy, **top**, are among nature's most elegant creations. About 15 billion stars make up this galaxy in the constellation Triangulum, 2.5 million light-years from Earth. Closer to home, at 23,000 light-years, is the Hercules Cluster, **bottom**, a vast swarm of more than a million stars. The Hercules Cluster is one of approximately 150 similar globular clusters that are satellites of our Milky Way Galaxy.*

galaxies eight billion light-years away. If the distance to the Andromeda Galaxy were reduced to the typical reading distance between your eyes and this book, the most remote galaxies known would be a mile away. Their images in a photograph (or, more likely, computer-enhanced images gathered by an electronic detector more sensitive than photographic emulsion) are mere smudges. Yet each wisp represents the combined light from billions of suns that has been on its way to Earth since before our planet or sun even existed.

Thousands of galaxies have been studied and catalogued, and millions have been photographed dusting the blackness of space, like snowflakes in a cosmic blizzard frozen in time. Billions of star cities are thought to exist, but the universe does not go on forever. It has finite bounds in space and time —about 15 billion years' worth—before which there was nothing. As we shall see, one of the great triumphs (or perhaps conceits) of modern astronomy is its ability to offer an explanation for the universe's origin and evolution from time zero to today.

Billions and Billions

Like nothing else in human existence, astronomy encompasses distances, ages and sizes so enormous that it is almost impossible to form a mental image of any of them. The numbers alone are enough to overwhelm the imagination. How big is a million or a billion? A million sheets of paper about the thickness of the pages of this book would form a tower as high as a 30-storey building, but a billion-page stack (1,000 million) would soar more than 10 times higher than Mount Everest. And a trillion pages? (One trillion is a million million, or 1,000 billion.) The pile would reach more than a quarter of the way to the moon.

Another way to think of it is that a million seconds is 12 days, but a billion seconds adds up to more than 31 years, and a trillion seconds is 300 centuries, longer than the history of civilization. Yet even with these analogies, it is often less than graphic to rely solely on huge numbers. Sometimes a better perspective is gained through examining a model of the cosmos.

Starting close to home, if the sun is reduced to the size of a ping-pong ball, Earth becomes a mote of dust eight feet from it, with a smaller speck, the moon, nestled beside it a quarter of an inch away. Jupiter is a pea 40 feet from the one-inch sun. A

The view toward our galaxy from the vicinity of the Hercules Cluster reveals a vista of sprawling spiral arms reaching out from a glowing yellow core where 100 billion stars reside. Astronomers think that globular clusters like the Hercules Cluster are among the universe's oldest collections of stars, their ageing red stars perhaps 14 billion years older than the throngs of young blue stars in the spiral arms.

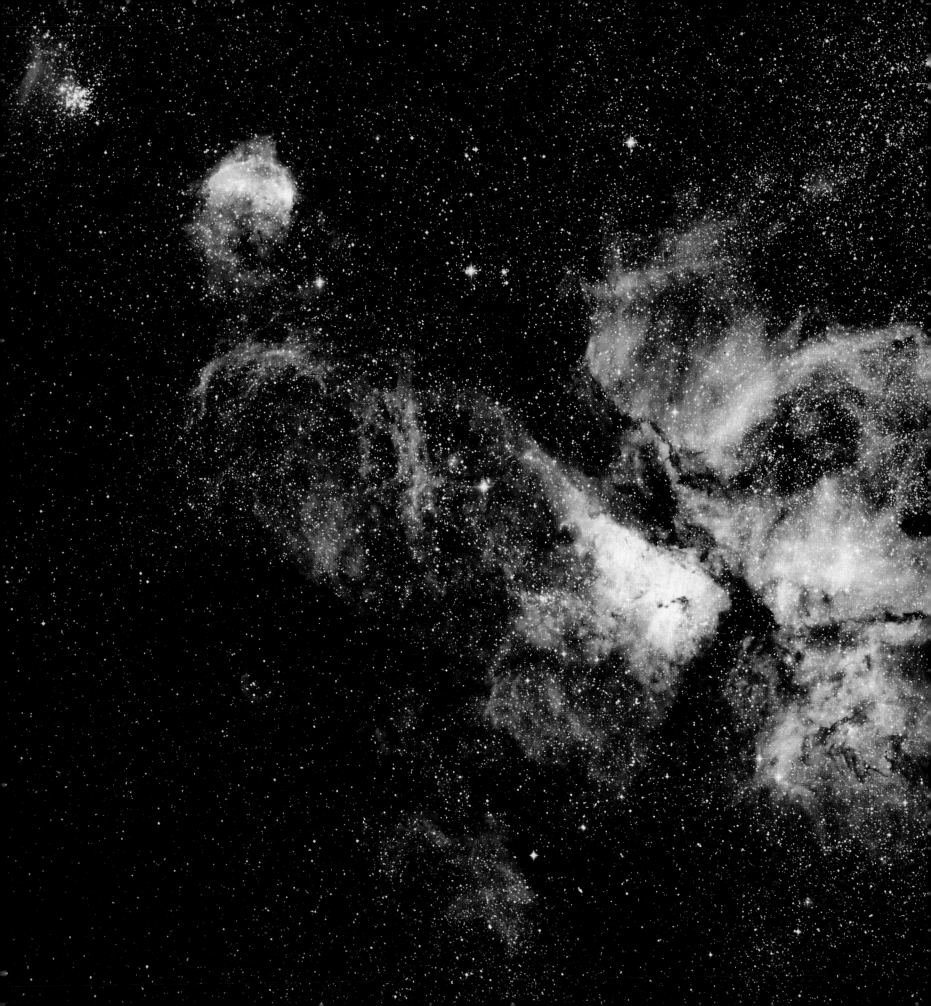

piece of dust 300 feet from the sun is Pluto, the outermost planet in the solar system, the sun's family. Comets are atomic particles, invisible to a microscope, extending in a cloud up to several dozen miles from the sun. Although there are trillions of comets, the vast volume of space they occupy keeps them, on average, several yards apart.

Alpha Centauri, the nearest star (a triple-star system), consists of two walnuts and a pea 400 miles away. Even if the universe were shrunk to this microscopic scale, it would be inconvenient to hike to the nearest star. If the model were encompassed in a volume of space the size of Earth, the vast hollow globe would contain only 800 stars, represented by walnuts, cherries, oranges, and so on. The billions of other stars in the galaxy would range well outside the Earth-sized region. Clearly, the universe is largely empty space.

To picture a larger portion of the universe, a further reduction of the model is necessary. The Earth's orbit is now the size of a dime. Most of the stars are dust-sized, and gigantic Betelgeuse, the largest star known, is less than one inch across. The average distance between stars has shrivelled to one mile, and the entire Milky Way Galaxy is as wide as the diameter of Earth. Our neighbouring galaxy, Andromeda, is a little over half the Earth-moon distance from us. If, like gods, we could stroll through our miniature universe, we would move from one star to another in only a few minutes. But the trek between galaxies would still take more than a lifetime.

Let us make one final reduction: the Milky Way Galaxy becomes small enough to hold in your hand, a delicate disc of subatomic-sized stars. The Andromeda Galaxy, only slightly larger, is not quite within reach at its scaled-down distance of seven feet. The universe of galaxies can be perceived in all directions, ranging from basketball-sized giant ellipticals to dwarf galaxies no larger than the head of a pin. Yet the immensity of the cosmos still defies the senses. There are so many galaxies in the universe, and they fill a volume of space so vast that even though they are only a few steps apart, it would take centuries simply to walk past all of these tiny islands of stars. And although we think it impossible to know what lies beyond our universe, there is no reason to presume that ours is the only one. The bubble of space we call the universe could be but one in an infinity of universes that constitute the true cosmos.

More than 200 light-years wide, the Eta Carinae Nebula is the largest known nebula (gas and dust cloud) in the Milky Way Galaxy. It lies 9,000 light-years away, buried in the spiral arm next inward from the sun's location. The Eta Carinae Nebula is one of thousands of star-forming regions in our galaxy where events similar to those that gave rise to the birth of our sun and its planets are now taking place. Pictures like this give the impression that stars are massed closely together in some regions of space. In reality, the stars here are several light-years apart, just as they are in the sun's vicinity. But because we are looking across an enormous volume of space thousands of light-years deep, an illusion of great density emerges in a two-dimensional image.

19

NEARBY WORLDS

CHAPTER TWO

F ive billion years ago, a dark, formless nebula —a hundred million cubic light-years of diffuse gas and dust—floated within the Milky Way Galaxy's graceful spiral arms. The vast cloud was one of thousands that populated the galaxy. While some of these dispersed, others would undergo an impressive transformation, a metamorphosis that was to be the fate of this particular cloud.

Barely perceptible at first, a pocket of the cloud began to contract, perhaps recoiling from the shock wave produced by a nearby supernova— the violent death of a massive star—or possibly as the result of a merging of several clouds. Whatever triggered it, the density of the collapsing region escalated when the contraction started; atoms once comfortably separated were jostled more vigorously, generating heat. In less than 100,000 years, nebular material at least a million times the Earth's mass had collected in a zone several times wider than the present orbit of Pluto, the most remote of the sun's planets. At its heart, a seething ball of hot gas, stoked by the crushing pressure of the infalling matter, reached the ignition point for nuclear reactions: the sun was born.

Meanwhile, the cloud material in the zone surrounding the primal sun had swirled into a disc, like a miniature spiral galaxy, called the solar nebula. The disc shape emerged because whatever internal motion the original cloud had was amplified during the contraction, just as a figure skater accelerates a spin by drawing in his or her arms. The process pulled much of the infalling material into the disc. Within the nebula, atoms and molecules, now swimming at closer range than in the near-

vacuum conditions of the initial cloud, began to combine into larger particles in somewhat the same way that ice crystals—the precursors of rain or snow—form in the upper atmosphere when air becomes saturated with water molecules.

The source cloud contained abundant hydrogen and helium, lots of oxygen, nitrogen, carbon and neon, moderate quantities of magnesium, silicon, iron and sulphur and lesser amounts of all the other elements. The crucial point is that the youthful sun heated the solar nebula, leaving only solid particles of elements that did not melt. The particles combined like adhering snowflakes over the next few million years, eventually accumulating into larger and larger bodies: planets, moons, asteroids and comets. Their compositions varied depending on their distance from the sun. A bull's-eye pattern emerged. Closer in, rocks and metals survived while almost everything else vaporized and ultimately dispersed (Mercury, Venus, Earth, the moon and Mars formed here). Farther out, carbon-rich composites similar to charcoal dominated, with rock and metal mixed in (the majority of the asteroids). More distant still, reddish materials rich in carbon-nitrogen-water compounds were common (outer asteroids and small moons of Jupiter). The sector of the solar nebula beyond this was cool enough for water-ice particles to remain along with hydrogen and helium gas (Jupiter and Saturn and their major moons). The most remote realm was frigid with ammonia, methane and carbon-monoxide snow (Uranus and beyond).

At the contracting cloud's nucleus, enormous quantities of heat generated by the rapid compres-

Mankind is made of star-stuff, ruled by universal laws.

HARLOW SHAPLEY
1962

The birth of the sun and its family of planets took place 4.6 billion years ago in a dark corner of a collapsing cloud of gas and dust that was to be the genesis nebula for hundreds of new stars. The incipient sun, a dull red ember in the black mass that extends from the centre to the left, is glowing from the heat of compression as the cloud material collapses under its own weight, adding to the sun's mass. A few dozen stars in another sector of the cloud at right have just come to life through ignition of their internal thermonuclear fires. Radiation pressure from the newborn stars peels back the nebular material to expose the cloud's interior. The red trail near the centre is a cometlike effect due to nebular material being blown back by stellar radiation from a dense knot.

21

*The crater-battered face of the moon, **above**, is a cosmic museum dating back to an era of violent collisions that marked the first half-billion years of the solar system's existence. A young Earth, **right**, was bombarded just as heavily as the moon, but aeons of weathering, erosion and mountain building have erased almost all traces of the primordial concussions.*

sion of nebular material as it rushed into the blazing star produced a far more luminous sun than we see today. The outward flood of energy soon reversed the flow of infalling matter, creating a billowing wind that swept away the gases of the solar nebula but left the solid particles. The bull's-eye configuration dictated that the solar system would emerge with differences in fundamental makeup.

The inner zone, the region of rock and metal, is the most familiar because we are in it, along with the moon, our companion in space, and the neighbouring worlds Mars, Venus and Mercury. Robots have probed and analyzed the surfaces of Mars and Venus and have gathered high-resolution pictures of Mercury from a flyby mission. But it is the Earth-moon system that we know best.

Earth and Moon: A Gravitational Embrace

Earth and the moon are a binary planet system. If the moon were circling the sun in its own orbit, say, between Earth and Mars, astronomers would not hesitate to classify it as a planet in its own right. Its diameter is two-thirds that of Mercury, significantly larger than Pluto and a respectable one-quarter of the Earth's diameter. Yet there it is, obediently swinging around Earth. None of the other terrestrial planets have anything resembling our moon. Venus and Mercury have none, and Mars's two Manhattan-sized moons are microscopic on a cosmic scale. Even when compared with the moons of the giant outer planets, our satellite is more than respectable. Jupiter's Ganymede and Saturn's Titan are the size of Mercury, but their planets are colossal relative to Earth. Pluto, which is so small that it hardly rates planet status, is endowed with a moon half its diameter. But Pluto is different in so many ways from all the other planets that it offers a weak basis for comparison.

Just how our moon was created and came to be orbiting Earth was a mystery until very recently. For more than a century, astronomers had debated the merits of three scenarios: the adopted-cousin theory (the moon was a small planet gravitationally captured by Earth); the sister theory (Earth and moon were born as a double planet); and the daughter theory (the moon fissioned from a rapidly spinning primordial Earth). Yet studies of the nearly one ton of lunar material returned by the Apollo astronauts failed to support any of these ideas. Instead,

a fourth hypothesis emerged and is now the frontrunner.

The new scenario—we'll call it the chip-off-the-old-block theory—is a product of 1980s' computer simulations of the formation of the solar system. The simulations suggest that 10 million years after the solar nebula initially evolved, the material in the region where Mercury, Venus, Earth and Mars were emerging had built up thousands of mountain-sized bodies called planetesimals. A few of these planetesimals might have been gargantuan, having up to three times the mass of Mars. The heat generated by colliding with something that big would have melted substantial portions of the nascent Earth. Debris from both Earth and the impacting body would have vaporized and splashed into nearby space. However, a portion of it could have lingered in the Earth's orbit, eventually coalescing into the moon.

The chip-off-the-old-block theory fits with the moon-rock analysis, which revealed that moon material is substantially different from Earth rock—

so much so that the sister and daughter theories became untenable. The moon contains very little iron and lacks such volatiles (more easily vaporized substances) as water, chlorine and potassium, which indicates that the moon was at one time heated to incandescence.

Melting caused by the planetesimal impact would account for the shortage of volatiles. The iron deficiency could also be explained by the impact, because the splashed-out material that formed the moon would have come from the surface of Earth and the planetesimal, rather than from their iron-rich cores. The composition of the entire moon seems to be more like the Earth's mantle than its interior.

The chip-off-the-old-block theory also explains the uniqueness of the Earth-moon system among our family of planets, because a big moon would result only from a small proportion of collisions early in the solar system's history. Therefore, it is not surprising that only one oversize moon exists. The adopted-cousin theory seems much less plau-

sible; computer simulations have consistently failed to re-create a reasonable capture sequence.

Although the moon is comparatively close in size to Earth, as satellites go, the surface conditions on the two worlds are completely different. Those now historic images of spacesuited astronauts bounding over dusty lunar plains tell much of the story. The moon is the ultimate desert: no air, no water. A grilled wasteland during the two-week lunar day, it is a freezer, almost as cold as deep space, for the two-week night. It all stems from having a relatively small mass, with its consequently weaker gravity (one-sixth Earth's). Any air or water the moon once had could not have lasted for more than a few million years. The simple heat of sunlight would give air and water molecules sufficient energy of motion to escape the weak lunar gravity and become dispersed into space.

Smaller mass also means reduced pressure of compaction on the interior of the moon, compared with Earth. Planet-wide modifications caused by volcanoes and earthquakes are drastically re-

Looming above the landscape of its inner moon Phobos, Mars is a world of contrasts: craters, valleys, sand dunes, ice fields and volcanoes. The huge volcano Olympus Mons is seen near the sunset terminator, its 70,000-foot summit creating a rare cloud on its leeward side. Phobos is one of the solar system's smallest moons—only as wide as the limits of a medium-sized city—and its gravity is so feeble that an astronaut with a strong throwing arm could put a baseball in orbit around it.

duced on the moon. It is a geologically quiet place: seismometers left by the Apollo astronauts registered vibrations from a baseball-sized meteorite striking the surface on the opposite side of the moon. Major surface-shaping forces such as plate tectonics (continental drift) and midocean ridges on Earth are completely absent on the moon. Its surface is a museum that records not internal but external processes. The craters are evidence of impacts from without—yawning bowls blasted during the solar system's exuberant youth.

Earth suffered the same bombardment. Indeed, as we have seen, the moon may have been born out of a particularly violent encounter with a planetesimal. Primitive Earth, cratered and without today's atmosphere and oceans, probably resembled the moon. Over time, steam and gas from volcanoes could have supplied our planet with water and air. More water would have arrived via impacting comets, which are largely composed of water ice. The atmosphere has evolved since then, but little escapes the Earth's gravitational grip. Atmospheric forces helped erase evidence of craters through weathering (wind, rain, rivers, glaciers), while the Earth's internal engine restructured the surface with plate tectonics, mountain building and volcanic activity. That, plus the Earth's temperature environment, which is dictated by its distance from the sun, has made it the only world in the solar system that has sustained large amounts of liquid water on its surface for billions of years. That in turn provided the crucible for life—a phenomenon which has been here for at least 3.5 billion years.

Mars: A Nice Place to Visit, but . . .

Mars is not an unfriendly place. In many ways, it is the most Earthlike planet we know about. The balmiest conditions on Mars resemble a summer day in Antarctica. But an explorer would need a spacesuit, since the air is not breathable. It is mostly carbon dioxide, and there is not much of it—its density is less than 1 percent of the Earth's atmosphere.

The main problem, however, is the cold. At noon on the hottest day of summer near the equator, the soil temperature might rise to 70 degrees F for a few hours; but the thin atmosphere never gets that warm. When the air temperature even approaches the freezing point of water, it is a Martian heat wave. In any case, the air pressure is too low for water to exist as a liquid. By sunset, both air and ground are as frigid as a winter day on Baffin Island. At midnight, any place in Antarctica would be warmer than Mars's equatorial zone. In the Martian polar regions in winter, the bottom really falls out, with daily *high* temperatures in the vicinity of 180 degrees F below zero. That is cold enough for the Martian air to freeze—and it does, producing a frosty blanket of dry ice (frozen CO_2) over the pole experiencing winter.

Subzero temperatures will not stop future Mars explorers, who will be outfitted in the fully heated and air-conditioned spacesuits that are available today. Getting humans to Mars and back is technologically possible, given the commitment and the funds. It would take almost three years for the round-trip, so the major hurdle is designing life-support systems for such long-duration voyages away from Earth-based supply ships.

In 1985, high-level space-programme officials in both the United States and the Soviet Union issued reports stating that landing humans on Mars was not only feasible but a reasonable goal for the early 21st century. Since then, the Soviet Union has ceased to exist, and interest in human exploration of Mars in both Russia and the United States has declined, at least as national or international enterprises. In 1960, science fiction author Arthur C. Clarke predicted that humans would walk on Mars before the end of the 20th century. It is now clear, that will not happen. Nevertheless, in the long view, the question is not if, but when.

On a summer afternoon in the Martian Amazonis desert (a likely landing site), an explorer would be struck by the colour—or lack of it. Everything is shades of orange, ranging from the bright peach-coloured sky to the rusty orange dust that covers the surface. The thin atmosphere filters little sunlight, so daylight is almost as strong as it is on Earth, although the sun appears half the size because of its increased distance. On the horizon, enormous sand dunes rise and fall in a procession of frozen waves. The ground is firm and littered with pebbles that more closely resemble cinders than smooth or granitic Earth rocks. Larger versions of those cindery rocks are scattered across the low, undulating landscape. No mountains or cliffs, at least not in this region. Certain sectors of Mars have extremely rugged terrain—canyons, volcanic flows, crater walls—while huge tracts are relatively flat.

A cloud of dust stirred up by a flick of the toe would be quickly dispersed by the sharp breeze.

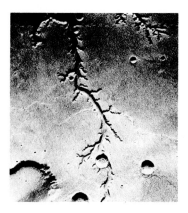

A mile-wide channel, almost certainly carved by water, scars a Martian plain that has not felt a drop of rain for billions of years. A flash flood from melted permafrost may account for this and hundreds of similar features, suggesting that Mars was once more Earthlike than it is today.

Mars, a planet once thought capable of supporting life, is largely a rock-strewn, wind-swept desert, according to spacecraft investigations. The planet has some Earthlike features—polar caps, morning mists, volcanoes—but it is distinctly unearthly in many ways, with its craters, global dust storms and lack of liquid water.

A typical landscape scene on Mars, **right**, was photographed in 1976 by the Viking 1 lander spacecraft, the first device to return pictures from the surface of the red planet. Parts of the spacecraft are seen in the foreground. The large boulder is about the size of a compact car.

Some features on Mars, such as the gigantic Mariner Valley, **bottom right** and illustration **facing page**—four times deeper than the Grand Canyon —will be natural targets for future exploration. The Mariner Valley scars the Martian land-scape for a distance equivalent to the width of North America.

Dust storms on Mars can be ferocious, with winds of more than 200 miles per hour whip-ping the dust particles high into the atmosphere. Since atmospheric dust never com-pletely settles, the sky is a peachy orange colour. The overall orange hue of the planet originates in the surface soil, which is iron-rich clay-like dirt with vast amounts of oxygen chemically locked into it. (The Viking spacecraft analyzed Martian soil and found it to be 50 percent oxygen.) Mars is basically a rusted world.

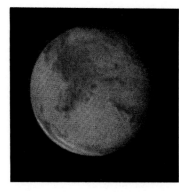

*Dwarfing any earthly geologic feature, the Martian volcano Olympus Mons, **right**, rises almost 90,000 feet above the surface of Mars, three times higher than Mount Everest, and sprawls over an area larger than France. The mountain contains enough rock to pave all of North America to a depth equal to the height of a 60-storey office tower. The caldera at the massive structure's summit is 50 miles wide and almost 9,000 feet deep. Geologists estimate that the most recent eruptions pushed enormous quantities of lava down the giant volcano's flanks less than 100 million years ago, suggesting that Olympus Mons is almost certainly not extinct. The virtual absence of craters indicates the entire region was geologically reworked around that time. **Above:** A Hubble Space Telescope image of Mars —the best shot of the planet ever obtained from the vicinity of Earth.*

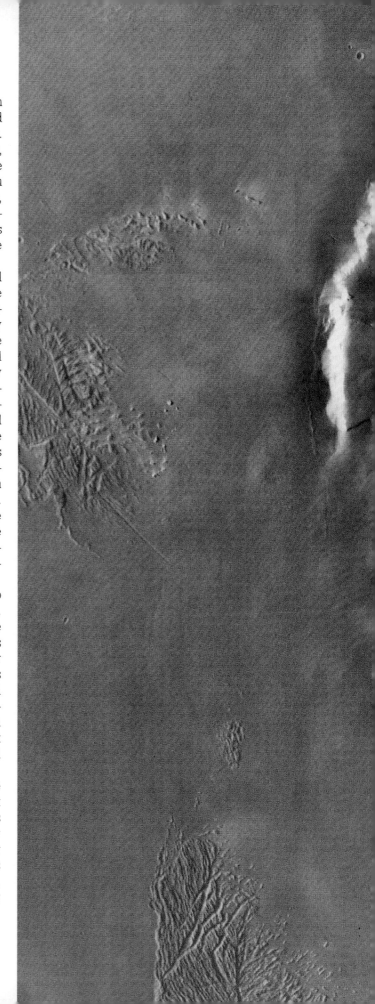

Summer is the windy season on Mars. Gusts can exceed 200 miles per hour, but with no exposed skin, a spacesuited explorer would not feel them. Nor would the astronaut be buffeted by the blasts, since the low-density air has far less force than the winds on Earth. However, fine dust particles from the orange desert are swept into the atmosphere, producing the peach-coloured sky. The strong seasonal winds kick up so much dust that the sky is never sufficiently free of it to expose the true hue of the Martian air—a deep bluish purple.

The colour of the Martian sky came as a total surprise to scientists when the first pictures of the surface of Mars were returned from the U.S. Viking 1 spacecraft that landed on the planet in July 1976. In the rush to get images out to the media, the initial digitally encoded colour pictures received by Mission Control at the Jet Propulsion Laboratory in Pasadena, California, were shifted in tone to produce a deep blue sky, which is what everyone expected. The processing gave the surface material a muddy greenish brown cast. But that was not the real Mars. By the following day, the researchers had decided to look more carefully at the colour-calibration data the spacecraft had provided, which showed that the sky was in fact pinkish orange. When the photographs were reprocessed, the Martian surface took on its familiar rusty orange colour. The incident exemplifies Earthlings' generally unconscious but pervasive tendency to transport terrestrial biases to other worlds.

True, Mars has some uncanny similarities to Earth. Its rotation period—its day—is 24.6 hours. The planet's axis, tipped at almost exactly the same angle as the Earth's, produces seasons analogous to ours, although because of the planet's larger orbit, a Martian year is 1.9 Earth years. But Mars was born both too small and too far from the sun. A larger planet with a correspondingly greater gravitational attraction might have been able to retain a denser atmosphere. Thicker air absorbs more heat from the sun, raising the surface temperature and allowing liquid water to exist.

At one time, Mars was probably much more like Earth than it is today. Dozens of channels that look like dry riverbeds extend over hundreds of miles of the planet's surface and were likely carved by flowing water. But if water ever existed as lakes or oceans sometime in the past, it was billions of years ago. The channels themselves are old and parched, unnourished for aeons. Some of that original water

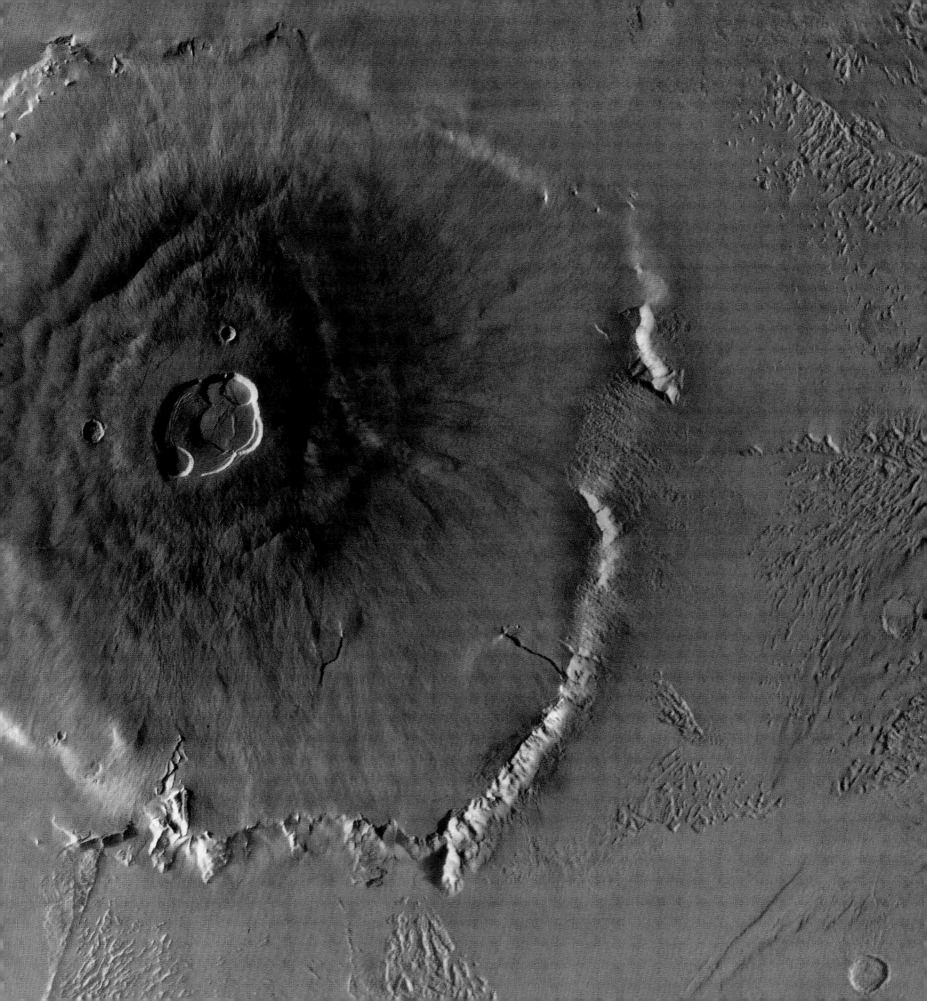

*Mars, **above**, is an arid world of windblown deserts, canyons and volcanoes, as seen in this equatorial view assembled from Viking orbiter images. The blank, nearly featureless face of Venus, **facing page**, is caused by a thick veil of haze, smog and clouds.*

nels, it was not long before the water evaporated into space or seeped underground and refroze. Under the conditions that exist today, a bowl of water placed on the surface of Mars would vaporize so quickly, it would literally explode. The atmosphere is starved for moisture. Yet if the red planet could somehow be warmed up, the permafrost melted and the atmosphere made a little denser, some scientists predict there could be large lakes, if not oceans, on Mars.

Is there life on Mars? The biological experiments carried out in the two Viking spacecraft that landed on the planet in 1976 yielded negative results. If life as we know it exists on Mars, it is not pervasive. Even the Earth's hardiest organisms would not survive. First, there is the cold; second, an almost total absence of liquid water; and third, a thin atmosphere that permits the sun's ultraviolet light to stream to the surface. Unattenuated solar ultraviolet light is highly destructive to biological organisms. (One method to purify well water utilizes a holding tank exposed to ultraviolet lamps.) Perhaps the most important legacy of our exploration of Mars is that our preconceived notions of a small, cool version of Earth have been dispelled.

Venus: Not a Nice Place to Visit

Venus is hell; at least, that is the message we have received from robot spacecraft which have visited the planet. Not only are the surface rocks of Venus hot enough to fry eggs, but the eggs would be vaporized. A human standing unprotected on the surface of Venus would be simultaneously sizzled by the heat, asphyxiated by carbon dioxide, crushed by the density of the planet's atmospheric cloak and scorched by hydrochloric-acid vapours.

Two French-designed balloon-borne experiments, dropped into Venus's atmosphere by the Soviet Vega spacecraft as they flew by the planet in 1985 on their way to Halley's Comet, found that Venus is just as hellish in its upper atmosphere as it is on the surface. Clouds and haze far above typical jetliner cruising altitude on Earth are laden with sulphuric-acid droplets as concentrated as car-battery liquid. Two surface landers, released at the same time as the balloons, descended by parachute to the scorching plains of Venus, where they drilled into the surface material and collected samples for examination. The results of the experiment indicate that the rocks and soil of Venus's sur-

is almost certainly still on Mars, but it must be in the form of ice—permafrost—frozen like cement below the surface, as it is in the Arctic regions on Earth. Another ice-storage locker is in the polar caps, which may be up to a mile thick (the frozen carbon dioxide is simply a frosting that covers the polar regions and extends beyond the water-ice polar cap during winter at each pole).

The channels indicate episodic and perhaps catastrophic releases of water on Mars, possibly due to heat from the interior's melting a region of permafrost and releasing a cargo of water that flooded the surface. Judging from the appearance of the chan-

face resemble those of Mars more closely than they do the Earth's. Thus the only world that we know to be similar to Earth in size, mass and surface gravity is a totally different place. Dreams of a tropical but not inhospitable world—so common in science fiction descriptions of Venus—are now shattered.

The first hint of the broiling surface conditions on our "sister" planet was revealed by radio telescopes in the 1950s. Confirmation came in 1962 from readings taken by the U.S. flyby probe Mariner 2. Because of an atmospheric greenhouse effect, temperatures on Venus from equator to poles exceed the melting point of lead. A series of Soviet surface landers, which followed in the years after Mariner 2, found that Venus's atmospheric blanket is 90 times as dense as the Earth's—similar in density to water at a depth of 3,000 feet—and 500 Fahrenheit degrees hotter than boiling oil.

The bad news about Venus was difficult to accept. In 1965, Cornell University astronomer Carl Sagan offered a brighter picture for future generations: Venus might be transformed into a more Earthlike environment through some form of global cultivation. "It is not inconceivable that an organism can be found or developed which will live and thrive somewhere on Venus and in time make it habitable for higher forms of life." Sagan proposed that the clouds of Venus could be seeded with microorganisms, which "would live a wholly aerial life suspended by the turbulence that stirs the clouds." The most likely choice for such an organism, Sagan speculated, is blue-green algae. The algae would transform Venus's atmosphere by extracting water and carbon dioxide, breaking them down through photosynthesis and discharging oxygen into the atmosphere.

Eventually, the surface would cool to a temperature that could support hearty plants and animals from Earth. "It may even become suitable for sustaining human colonization," suggested Sagan. This vision of terraforming Venus caught on. Dozens of writers repeated the same theory, postulating that the transformation of Venus to Earthlike conditions might take as little as a few centuries.

Could it work? The scenario was analyzed in the early 1980s by James Oberg of the National Aeronautics and Space Administration, who found that the project would be a herculean effort far beyond our present or projected technology. Oberg also concluded that simple algae seeding would not work. "Assuming we want to intervene in nature,"

31

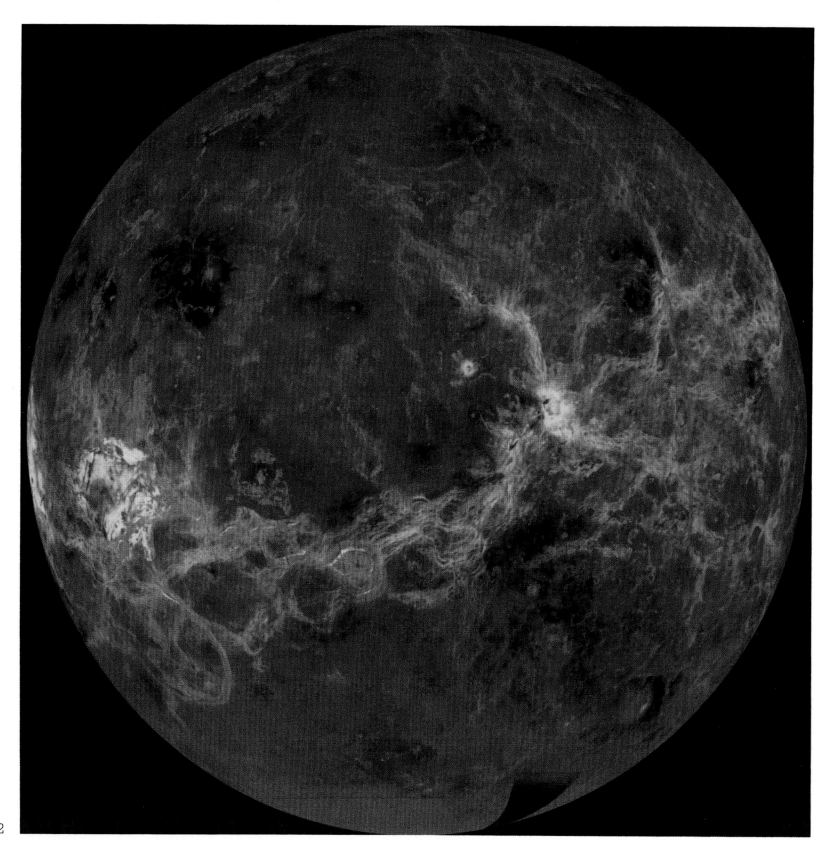

he said, "there are major roadblocks such as the spin of Venus, which is too slow for comfort." One day on Venus is equivalent to 59 Earth days and is followed by an equally long night. Life would be impossible on Earth if it rotated that slowly because of the enormous heat buildup during the day and the extended nightly deep-freeze. Capturing and flinging comets into our sister planet has been suggested, since they could provide water as well as accelerate Venus's rotation (if properly targeted); however, according to Oberg, the momentum involved in such collisions would not significantly alter the spin rate of Venus. Furthermore, the heat generated by such impacts would be additional excess energy that would have to be dissipated to bring the planet to Earthlike temperatures.

Even if the stifling atmosphere could be removed and the rotation problem overcome immediately, Oberg reasoned, the hot surface rock would require centuries to cool from a searing 860 degrees F. Venus's lack of a natural magnetic field (because of the planet's sluggish rotation, which does not activate an internal dynamo like the Earth's) may have as-yet-undetermined detrimental effects on biological activities. The role our magnetic field plays in protecting life is only vaguely understood. If all the obstacles could be surmounted and Venus were successfully transformed into an Earthlike environment, the favourable conditions would soon deteriorate because the intensity of sunlight that caused the original situation would be constantly acting to reverse the process.

Why So Unlike Earth?

The swamps of Venus and the canals of Mars, so often described in science fiction, were two major casualties of the space age. They were wonderful dreams that lasted through the 1950s when the question of life on other worlds became a part of popular culture—a spin-off from the then new UFO phenomenon and Hollywood's heyday with cheap black-and-white science fiction movies. About the same time, a concept dubbed the "ecosphere" merged into astronomical thought. The ecosphere is a region around the sun and other stars where a planet similar to Earth would have surface temperatures between the freezing and boiling points of water—a star's habitable region, a place in front of the stellar hearth where life as we know it would be neither frozen nor sizzled.

When the ecosphere is applied to our solar system, it extends from just inside Venus's orbit to beyond the orbit of Mars, and Earth wheels around the zone's midsection. With three out of nine planets in the comfort zone of our star system, the idea took root that every star similar to the sun should have at least one planet tucked comfortably inside its ecosphere. The concept received "royal assent" by being favourably described in an influential article by astronomer Su-Shu Huang in *American Scientist* magazine in 1959.

Two significant notions emerged from this thinking: first, that life-bearing planets ought to be commonplace in the universe; second, that Mars and Venus should be good candidates for harbouring some form of life. I don't think it is any coincidence that at about the same time, the first articles on methods of detecting radio communication from extraterrestrials were published in highly respected scientific journals (see Chapter 8). The problem—and this is nothing new—is that the theory was naively Earth-centred.

Some of the most important scientific concepts in astronomy are elegantly simple and easily understood, and the ecosphere idea seemed to be one of these. But it turned out to be hopelessly simplistic, since it was primarily based on our meagre knowledge in the 1950s—hardly better than guesswork —of the surface conditions on Venus and Mars. We now know what happens when a planet initially almost identical to Earth is positioned at Venus's distance from the sun. The result is not a tropical Earth but the hellish environment of Venus.

It is impossible to say just how close Earth could be to the sun before reaching the point where the runaway greenhouse effect would take over and the transition from Earthlike to Venuslike conditions would occur. But obviously, being in the ecosphere did not do Venus much good—nor Mars, with its water supply frozen as permafrost and trapped in polar icecaps. When the biology experiments aboard the two Viking landers failed to detect life as we know it, the century of wishful thinking following the "discovery" of the Martian canals faded forever. Geocentrism thrives on speculation but withers in the face of hard data.

The more we learn about the environments on our neighbour worlds and the evolutionary history of the solar system, the more the ecosphere sinks into fantasy. Determining how far a life-supporting planet "should" be from its star is complicated

Global view of Venus, **facing page**, *shows the planet as if it had no atmosphere. This unveiling of Venus is mainly the work of the radar system on board the Venus-orbiting Magellan spacecraft,* **above**. *It reveals a surface of faults and lava flows, with few craters. Lighter areas are the roughest surfaces. Scientists estimate that the entire planet was resurfaced by lava about 800 million years ago, perhaps in a catastrophic episode. There is no evidence of tectonic plate motion, or continental drift, like that on Earth, and no active volcanoes have been detected. The yellow-brown surface coloration is what an explorer would see in normal sunlight filtered through the dense atmospheric haze.*

33

Bathed in the dull glow of daylight filtered by a suffocatingly dense atmosphere, the Soviet spacecraft Venera 14, **top***, gathered valuable information about the surface and atmosphere of Venus in 1982. Three years later, the Soviet Vega 2 lander descended to the planet's night side,* **bottom***, where in this rendering, it is seen temporarily illuminated by distant bolts of lightning. The 860-degree-F surface temperature sizzled the well-protected spacecraft, and both ceased to function after an hour in the hellish environment.*

by a plethora of factors. In the case of Mars, for example, we are dealing with a body about half the size of Earth but only a tenth the mass. Worlds less massive than our planet have trouble hanging on to an atmosphere, since lower gravity means that more atoms and molecules of air escape into space. Atmospheric escape from these less massive planets is further accelerated by solar radiation. But most important is unequivocal evidence gathered in the last two decades which shows that atmospheres evolve over time.

Another ingredient in the equation comes from astrophysics. The sun, it seems, has almost doubled in brightness since shortly after its birth 4.6 billion years ago, when it stabilized into a hydrogen-burning star known as a main-sequence star. At that time, it was shedding light at about 60 percent of its present level. This has slowly but steadily elevated to the radiative output we enjoy today.

So the situation changes with time. A planet that experiences optimal conditions at a certain distance from its parent star might become sterile after a few billion years as the star cranks up the thermostat during its lifetime. Conversely, a planet gripped in icy cold may be thawed out as the warming star ages and evolves. But according to the fossil record, life has been abundant on Earth for at least 3.5 billion years of its 4.6-billion-year history, from a time when the sun was about two-thirds its present luminosity. Earth's average distance from the sun could not have changed, since its orbit is extremely stable. So how did our planet initially avoid a global ice age?

The answer appears to be planetary and atmospheric evolution. Now that more than $2 billion worth of sensitive space probes have landed on Venus and Mars, enough detail is known of their surface conditions for scientists to piece together a picture of planetary evolution which may explain how Venus, Earth and Mars came to be so different.

Clearly, the concept of a fixed ecosphere, or habitable zone, around a star must be discarded, both because our sun's luminosity has increased and because a planet itself can evolve over time—especially the insulating properties of its air, which change with variations in atmosphere composition. Atmospheres evolve through volcanic activity, through sunlight decomposing various atmospheric molecules and through the escape of certain gases from the gravitational pull of the planet. In addition, there is a mixing of gases with the surface rocks and

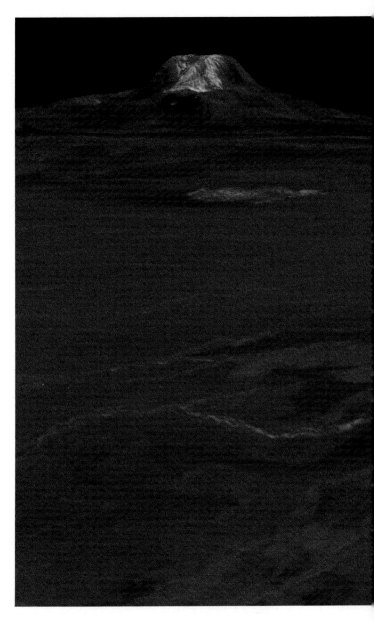

liquids. The picture is far more complex than ecosphere proponents ever imagined.

Today, the surface of Venus is as hot as the inside of a self-cleaning oven. But in the early days of the solar system, when the sun was dimmer, Venus's surface may have been cool enough for water to flow in rivers and to form oceans. Support for this scenario emerged in 1978 when the U.S. Pioneer Venus spacecraft sent several probes into the planet's atmosphere. Analyses of their measurements revealed that there are residual gases in the atmosphere which would be most easily explained by the existence of a global ocean of water for sev-

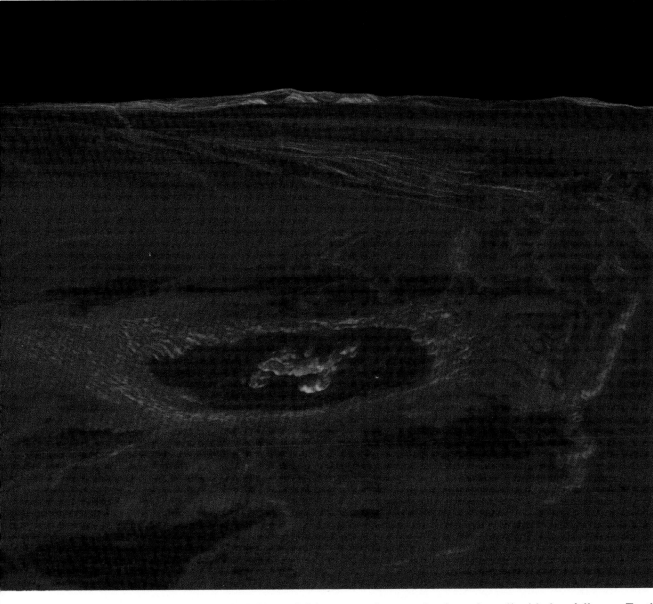

This dramatic view across the surface of Venus from the altitude of a low-flying airplane was computer-generated using Magellan radar data. It shows real features and approximates the daytime colour on Venus in the cloud-filtered sunlight, but the vertical scale has been exaggerated. The slopes of the 10,000-foot-high volcano Gula Mons on the horizon are much gentler than seen here. In general, Venus is flatter and less mountainous than Earth. The black sky is false, but it provides an easily distinguished horizon; the real Venus day sky is an unbroken bright yellow curtain of haze. Foreground crater is Cunitz, named for astronomer Maria Cunitz. All Venusian features are named after females.

eral hundred million years early in Venus's history. With the sun about two-thirds its present brightness, conditions on Venus might have been similar to those on Earth now.

Whether life got a toehold on Venus at that time is unknown, but as solar radiation slowly increased to present levels, the Venus ocean was doomed. Eventually, the ocean began to boil, filling the atmosphere with water vapour. Recent theoretical work on the composition of the original atmospheres of the inner planets suggests that large amounts of carbon dioxide were present on Venus, Earth and Mars. Since those early days, the level

of atmospheric carbon dioxide has fallen on Earth (today, one-tenth of 1 percent of the Earth's atmosphere is carbon dioxide) but not on Venus. Active evaporation of the primordial ocean on Venus added water vapour to the carbon dioxide in its atmosphere. Carbon dioxide alone creates an atmospheric blanket that acts like the roof and walls of a greenhouse, trapping infrared radiation. Water vapour is even more efficient at producing the greenhouse effect, heating the planet further.

As the temperature soared on Venus, the gradually escalating solar input added fuel to the flames, heating the surface to levels that melted

Details of Mercury's cratered surface were provided by the cameras of the U.S. Mariner 10 space probe that swung by the innermost planet in 1974. No other spacecraft has passed near Mercury, and there are no plans for further exploration of the desolate world until well into the 21st century.

some of the rocks, which sent more carbon dioxide into the air and continued to drive up the temperature. Volcanoes introduced additional carbon dioxide from Venus's interior along with other gases, and the greenhouse cycle spiralled until the planet reached its present temperature, which exceeds the melting point of lead. Today, Venus's atmosphere contains about 300,000 times as much carbon dioxide as Earth's.

Surprisingly, the haze and cloud decks in Venus's upper atmosphere reflect sunlight more effectively than the Earth's clouds do, so the amount of sunlight reaching the surface of Venus is less than Earth receives on a cloudy day. Hypothetically, if the clouds could be retained but the greenhouse effect eliminated, Venus would be cooler than Earth. This is wishful thinking, though. Venus is trapped. With the sun continuing to warm and no liquid water to cycle Venus's carbon dioxide back into carbonate rocks, the planet is a smothered inferno.

On Earth, however, the initial elevated level of carbon dioxide was exactly what was needed to keep our planet from becoming locked in a worldwide ice age. The key to the Earth's survival was that it never became warm enough for the surface water to start boiling. The oceans contain more than 50 times as much carbon as the atmosphere, mostly dissolved in the form of bicarbonates. If the oceans ever did boil, the oxygen in the water would combine with the carbon to form carbon dioxide in huge quantities—as must have happened on Venus—and the hydrogen would escape into space.

Apart from the essential direct role water plays in life processes, oceans keep carbon dioxide from building up by absorbing it from the atmosphere and then cleansing it from the water through the production of carbonate rocks and seafloor sediments. Rain, which contains dissolved carbon dioxide from the atmosphere, aids the cycle when it enters the oceans. Once there, carbon is taken up by marine organisms. When the organisms die, their shells sink to the bottom and form limestone. The reverse process occurs when a volcano spews carbon dioxide formerly trapped in the rocks back into the atmosphere or into the oceans from the volcanic midocean ridges.

This can be a self-regulating process. As the sun gets hotter, it evaporates more water from the oceans, which causes more rain, thus reducing the atmospheric carbon-dioxide content. When green plants appear, they remove more carbon dioxide from the air and oceans and return oxygen—nowhere near as effective an insulating atmospheric gas as carbon dioxide. Warm and cool cycles in the Earth's climatic history may be directly related to carbon-dioxide reduction. These in turn can be traced to plate tectonics. When plate motion is most dynamic, more carbon dioxide is introduced into the atmosphere; the temperature then rises, producing climate regimes such as the lengthy warm spell about 100 million years ago that marked the height of the Age of Dinosaurs, when most of the planet experienced tropical conditions.

In about four billion years, the sun will release 50 percent more energy than it now does as a result of a slight rearrangement of its internal heat-producing mechanisms. Mars will then receive enough solar heat to begin to thaw the global deep-freeze. It is a long time to wait, but when it happens, the permafrost will melt and liquid water will flow once again on the surface of Mars. The more Earthlike environment could last for millions of years, just as it probably did on Venus four billion years ago, and might allow some form of life to arise on the future Mars, but that is sheer conjecture. What seems more definite is that if Earth and Mars were to change places today, the Earth's oceans would freeze, and the planet would be locked in a global ice age that would not eliminate all life but would certainly reduce its range of environments.

Conversely, if Mars were cruising in the Earth's orbit, the permafrost in the equatorial region would melt. Whether the added atmospheric water vapour would be adequate to offer life-support environments is not known. But this kind of speculation is littered with caveats because we simply do not understand the conditions necessary for life's genesis nor the branching paths that life will take once it gets a foothold on a world. We have only one example: Earth.

Mercury: Sun-Scorched World

An astronaut trained for exploration of the moon would be well prepared for the rigours of the innermost planet of the solar system. Mercury is an airless, cratered world half again as large as the moon but still only one-third the diameter of Earth. Astronauts on Mercury would feel about twice the moon's gravitational pull but only a third of that experienced on Earth. Efficient spacesuit air-conditioning will be a prerequisite, with the sun shining six times

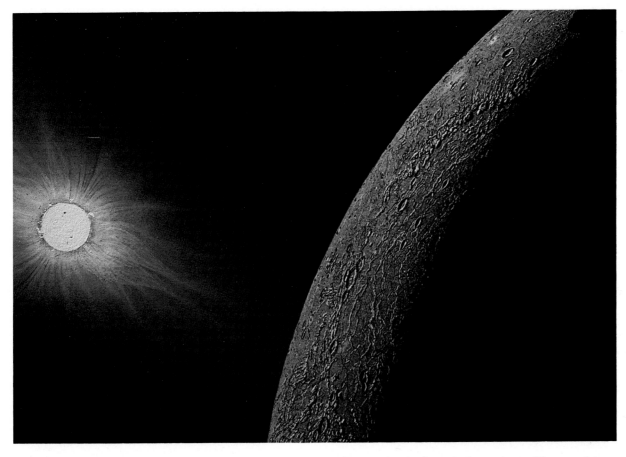

as intensely as it does at noon on Earth.

Of all the worlds in the solar system explored so far, Mercury proved to be the least surprising when the Mariner 10 spacecraft provided our first (and only) close-up views in 1974. Mercury's face, like the moon's, still shows the scars of saturation bombardment by asteroids and comets during the solar system's youth. For the last three billion years at least, it has remained essentially unaltered. The biggest changes have come in our perceptions of Mercury as exploration advanced.

The planet's proximity to the sun and its small size made it exceedingly difficult to study when the only available tools were Earth-based telescopes. Just a few pale smudges have been seen from Earth, even with the most powerful telescopes. Until two decades ago, textbooks stated as fact that one side of Mercury perpetually faced the sun, leaving a twilight zone around the sunrise-sunset transition, where some optimists suggested life might be found. Radar signals transmitted from Earth and bounced off Mercury in 1965 revealed the truth: the planet is turning, but slowly. It orbits the sun twice while rotating on its axis three times. The combination of these two motions results in a true day (sunrise to sunrise) that is 176 Earth days long. Noon temperatures on the day side reach about 700 degrees F, but because there is no insulating atmosphere, a spacesuit air conditioner could easily radiate into the vacuum of space the appropriate amount of absorbed heat to keep an astronaut cool.

Is Mercury worth exploring when the moon, a similar object, is hundreds of times closer? No return missions to Mercury are planned for the rest of the century. The planet is regarded as a low-priority target, although someday it may prove enticing. Mercury is believed to be the most mineral-rich major body in the solar system. However, I doubt if this will ever be used as a reason to go to the planet. If a pure mining expedition is ever contemplated, billions of tons of minerals, in the form of passing asteroids, fly much closer to us. Instead, as the inevitable bubble of exploration expands in the centuries to come, Mercury will have to take its turn among hundreds of other bodies in the solar system that will become targets of study.

CHAPTER THREE

REALM OF THE GIANTS

Hurricane-force winds whip up storms that could swallow the continent of Asia. Lightning bolts capable of vaporizing a small city rip open the sky. A swirling vortex of clouds larger than Earth has been raging for centuries and shows no signs of abating. Above it all, lethal radiation permeates the surrounding environment. This is Jupiter, a bizarre colossus that dwarfs every standard of comparison familiar to Earthlings.

Jupiter is the largest and nearest of four giant planets, all totally different in almost every respect from Earth and its companions in the inner solar system. The contrast between the planets close to the sun and those farther out can be traced back to the solar nebula.

Compared with the birth of the terrestrial planets, from Mercury to Mars, the formation of the giant planets in the outer solar system was a leisurely process taking perhaps half a billion years. One reason for their prolonged development is that objects move more slowly the farther they are from the sun, and the solar nebula was almost certainly less dense at Jupiter's or Saturn's distance than it was closer to the newly emerged star. But of greater importance, when the dazzling primal sun briefly shone with more than 100 times the present-day solar luminosity, it radically changed the environment in the inner solar system but only marginally affected the conditions from Jupiter outward. The composition of Jupiter and Saturn—four-fifths hydrogen, one-fifth helium—is basically the same ratio as in nebulas in the Milky Way where stars are being born today. This indicates these worlds coalesced from essentially pure solar-nebula con-

struction material, not the heavy, hard-to-melt residuals that make up Earth and its neighbours.

Jupiter is the best known and, due to its immense size, the most alien of the giant planets. Its visible surface is an endless maelstrom of clouds so vast that if Earth were peeled like an orange, the skin would not quite cover the Great Red Spot, the planet's largest storm centre. In size, this planetary giant is to Earth what a basketball is to a golf ball. Jupiter has no rivals in the solar system; it is more than twice as massive as all the other planets and their satellites combined.

The big planet's quilt of ivory- and salmon-coloured clouds is a facade, an ever-changing mask of many hues that blankets the real planet underneath, which is essentially a giant drop of liquid hydrogen and helium 318 times the Earth's mass. Encased at the very centre is a small rock-and-metal core approximately the size of our planet.

Seen from Earth, Jupiter is the fourth brightest object in the sky after the sun, the moon and Venus. Ponderously navigating its huge 12-year orbit about the sun at five times the Earth's distance, Jupiter drags with it a miniature solar system of at least 16 moons, two of which are the size of the planet Mercury. The giant planet completes one rotation on its axis every 10 hours, making its day the shortest of the nine planets. A point on Jupiter's equator races along at 22,000 miles per hour compared with 1,000 miles per hour for a similar point on Earth. The tremendous rotational speed combined with the fluid character of the planet make Jupiter bulge at its equator, giving it a polar diameter about 7 percent smaller than its equatorial diameter.

The atmospheres of celestial bodies as well as whirling cosmic nebulae can be regarded as the timeless sanctuary of animate forms.

JUSTUS VON LIEBIG
1861

Its nuclear furnace newly ignited, the youthful sun flared up and, in the process, pumped a powerful surge of radiation past the inner planets, sweeping the zone clear of gas and dust. The outer planets, protected by their distance from the sun, were largely unaffected by this phase, which lasted about 10 million years and left rocky inner worlds, such as Earth, and allowed the outer planets like Jupiter to retain massive gaseous atmospheres.

THE HUBBLE SPACE TELESCOPE

Everybody knows that the Hubble Space Telescope's optics were botched. The painful mistake was caused by improper installation of a small rod in the precision rack used to test the telescope's main mirror during its manufacture. This moved the test apparatus 1.6 millimetres out of position —about the thickness of a quarter—which is a huge error in the world of precision optics. The mirror was polished to a perfect optical figure, but the wrong one.

The flaw was discovered during routine tests shortly after the telescope was placed in orbit. One astronomer later moaned that the error in the mirror could have been detected by an amateur telescope maker using less than $50 worth of test equipment. But NASA, to its lasting regret, relied on a single high-tech testing procedure. No one expected that the test equipment itself would be assembled wrong. And no one was more disappointed to learn this saga of woe than the astronomers scheduled to use the orbiting observatory. Some have postponed their observations until after 1994, when space-walking astronauts are scheduled to give most of the instruments aboard Hubble corrective "eyeglasses" near the focus to compensate for the 94-inch main mirror's improper curvature.

But Hubble is far from blind, as the images here and on pages 28, 46 and 102-3 reveal. Even in its hobbled state, it is

more effective at most astronomical tasks than any telescope on Earth. Once the optical flaw (known to opticians as spherical aberration) was discovered, engineers and astronomers at the Space Telescope Science Institute in Baltimore worked to find ways of electronically subtracting the unfocused light (85 percent of the total) to sharpen the images. That process is most effective on bright objects, where removing 85 percent of the light is essentially irrelevant. The result is by far the best image of Jupiter, **top**, *and Saturn,* **bottom**, *ever obtained from the Earth's vicinity.*

The photograph, **facing page**, *shows the telescope being hoisted into space from the cargo bay of the shuttle Atlantis in April 1990 by the Canadarm remote manipulator system. The reflective door over the front aperture shields the telescope from exposure to the sun or to sunlight reflected off the Earth, either of which could damage the instrument's sensitive detector array.*

Although the telescope has made useful contributions to our knowledge of objects in our galaxy and nearby galaxies (pages 102-3), it has so far only scratched the surface of its original prime objective: refining the distance scale of the universe (see page 72). The targets in this category of research are usually faint stars in remote galaxies. Having to remove 85 percent of the light to attain sharp images means the exposures must be six times longer than if the optics were performing to specifications. Most of this type of exploration is awaiting the 1994 space shuttle repair mission.

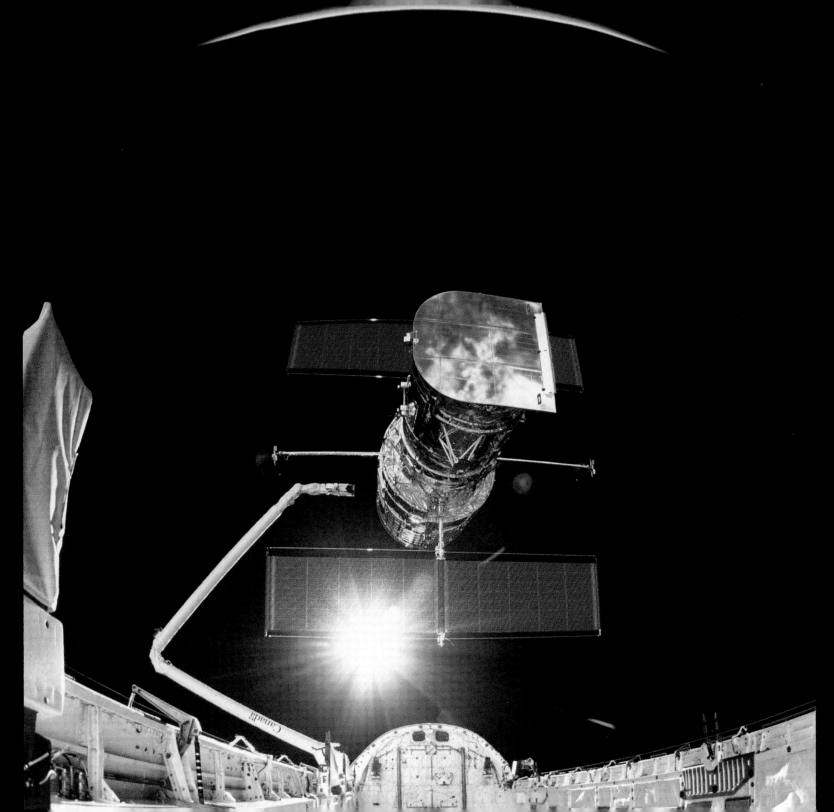

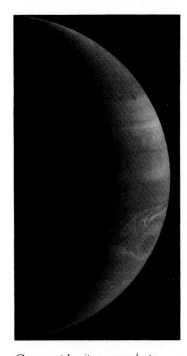

Crescent Jupiter was photographed by Voyager 2 in 1979 after the spacecraft had passed behind the planet's night side and begun a two-year trek to Saturn. Although a solid, rocky body about the size of Earth likely exists at Jupiter's core, **facing page**, *it could never be explored. The overlying layers of hot, dense hydrogen would crush any conceivable robot probe. The first human artifact to reach Jupiter's clouds will be the Galileo descent module scheduled to arrive in the mid-1990s. After radioing back measurements for the first few hundred miles of the descent, the probe will be crumpled by the pressure.*

Human exploration in the immediate environment of Jupiter seems improbable in the foreseeable future. The planet's intense magnetic field moulds deadly radiation belts into an invisible shield around it. The voyage would be almost as hazardous as an exploration of the sun; however, it may someday be possible for humans to probe the colossus using a shielded submarine-type device.

Approaching Jupiter's ocean of clouds, the crew of such a vehicle would descend first through rarefied haze layers of ethane and acetylene, then into the clear hydrogen-and-helium atmosphere with its tiers of streaming white clouds similar in appearance to cirrus clouds but largely made of ammonia-ice crystals. These clouds become thicker and more billowy as the descent continues, until nothing is visible but white sheets swirled by wind vortexes and occasionally skewered by cloud towers from below. At this level, an instrumented exploration probe suspended by a balloon inflated with heated hydrogen could ride the winds indefinitely. The temperature here is minus 184 degrees F, and the atmospheric pressure is about 70 percent of that on the Earth's surface. Eight miles above, at the 30 percent atmospheric-pressure level, the temperature is minus 230 degrees F.

Twelve miles below the tops of the white ammonia cloud decks is a transition zone leading to multicoloured ammonium-hydrosulphide-crystal clouds: beige, peach, khaki, salmon and brown. They form horizontal belts around the planet that alternate with the higher zones of ammonia clouds and constitute the two main features of the planet's visible face. These relatively permanent coloured belts and white zones are the planet's weather system, similar in some ways to the Earth's. But in the Earth's case, the weather machine is powered by the sun, whereas on Jupiter, the system is driven more by internal planetary heat than by solar radiation. On Earth, huge masses of warm, light gas rise to high altitudes, cool off, get heavier and then roll down the sides of new rising columns of gas. The force of the Earth's rotation creates an overriding west-to-east flow. Instabilities convert the motion into the enormous spiral storm centres (cyclones and anticyclones) seen in satellite photographs. The Earth's weather system is further complicated because the atmospheric flow is disturbed when air encounters continents and mountain ranges.

On Jupiter, the planet's more rapid rotation virtually eliminates north-south flow, whipping the clouds into distinct alternating high- and low-pressure bands parallel to the equator. Spiral storms are quickly twirled into continent-sized oval disturbances that turn like ball bearings between the main atmospheric bands. The biggest of these storms, the Great Red Spot, is the highest feature on Jupiter and is composed of material dredged up by an especially powerful cyclonic system. How it has sustained itself for so long is a mystery (drawings made in 1664 show it), but without landmasses to bump into, it is difficult for such a large storm to dissipate once it gets going. Because it intrudes well into the two counterflowing currents in the atmospheric wind zones to its north and south, the Red Spot may tap their energy of motion more effectively than a smaller vortex could.

Not far below the ammonium-hydrosulphide clouds, the temperature is high enough that ammonia is no longer frozen, and water-ice crystals drift around in the weather circulation along with ammonia droplets. A few miles farther down, the temperature rises above the freezing point of water. Frost and ice accumulation on the exterior of an exploration vehicle would melt at this level, where water is abundant in the form of alkaline droplets containing ammonia in solution. Someday, this zone will be thoroughly investigated because the ammonium-hydroxide mist here was once thought to be a lethal environment for all forms of life, but bacteria have been found on Earth that survive in hot springs far more alkaline than was considered possible only a few years ago. Other bacteria can withstand temperatures higher than the boiling point of water, and some do not require oxygen. There is no doubt these simple forms of life thrive on Earth under conditions that are normal for Jupiter. The question is: does life exist in Jupiter's atmosphere? Recent evidence suggests it does not.

Though not intended to answer the life question, the four Pioneer and Voyager probes that sailed by the giant planet during the 1970s provided an unprecedented examination of the eddies and currents in the Jovian atmosphere. Strong vertical motion seems to be prevalent, even violent, in some regions. The constant stirring could, within hours, carry atmosphere-borne life from the frigid upper cloud decks to the hotter lower realms—a temperature range of hundreds of Fahrenheit degrees. Although this does not eliminate the possibility of life in Jupiter's atmosphere, it is a discouraging sign, suggesting that the Jovian clouds may be too turbulent for even the most resilient known bacterium. A much better reading of atmospheric dynamics and specific temperature regimes will be provided by the U.S. Galileo atmospheric probe, a capsule that will descend by parachute directly into the Jovian atmosphere in the 1990s. The Galileo probe will return data until it is crushed by atmospheric pressure less than a hundred miles down.

The ammonium-hydrosulphide clouds form a ceiling above the potential life zone, where the atmosphere is misty but essentially clear. Below looms an abyss leading to total blackness. From this level down, our exploration takes on the characteristics of a deep-sea dive, as the atmospheric pressure escalates to more than 100 times the surface atmospheric pressure on Earth and sunlight is dimmed to near-nightfall levels. The pressure increase continues until the equivalent of seawater pressure in the deepest ocean trenches on Earth is crushing around our space capsule. The descent is eventually stopped 600 miles down by rising currents of liquid hydrogen denser than lead. At 3,600 degrees F, it is hotter than the inside of a blast furnace. The temperature at a depth of 1,800 miles is 10,000 degrees F, and the pressure is 90,000 times the atmospheric pressure on the Earth's surface.

About one-third of the way to Jupiter's core, the temperature reaches 20,000 degrees F, and the pressure is three million Earth atmospheres. At this level, liquid hydrogen turns into liquid metallic hydrogen, which has the properties of metal and is responsible for Jupiter's enormously powerful magnetic field. Deep inside Jupiter, near the core, incredible pressures are concentrated on a molten region where the temperature is 54,000 degrees F, five times hotter than the surface of the sun. Yet this does not even begin to approach the several million Fahrenheit degrees required to ignite the thermonuclear "fires" that power stars.

Jupiter has been called the star that failed, which suggests that it just missed achieving stellar status. While there are some similarities (Jupiter is composed of basically the same ingredients as a star), it is as far from being a star as Earth is from being a giant gaseous planet. To become a true star, Jupiter would have to be 80 times more massive to fire up its nuclear furnace permanently. Nevertheless, the planet's core is hot enough to generate seething convective currents that, in about a century, transport heat from the interior to the upper atmospheric levels, where the rise and fall of cloud currents

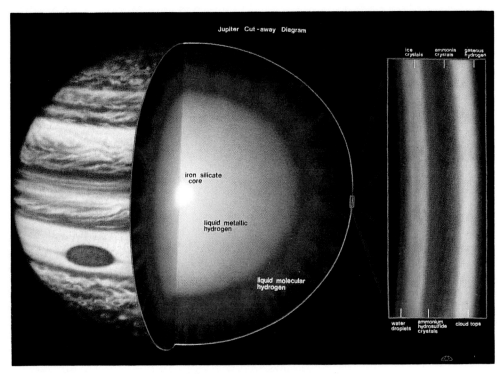

Jupiter Cut-away Diagram

iron silicate core

liquid metallic hydrogen

liquid molecular hydrogen

ice crystals   ammonia crystals   gaseous hydrogen

water droplets   ammonium hydrosulfide crystals   cloud tops

finally disperse it into space. The energy released is substantial: Jupiter radiates 2½ times as much heat as it receives from the sun. The best explanation for the existence of the high internal temperature is that it is maintained by a slow contraction of the planet. With such a huge mass, a small contraction goes a long way. A shrinkage of just a fraction of an inch per year can account for the planet's tremendous internal furnace.

Above Jupiter's highest cloud deck, the environment is just as unearthly as it is below. A deadly invisible shield—Jupiter's magnetic field—surrounds the planetary colossus. The total energy of the Jovian magnetic field is 400 million times the Earth's comparatively puny field. Jupiter's radiation belts, embedded in the field, are up to 100 times stronger than the dose lethal to humans and are near the tolerance limit for spacecraft systems. No other planet comes close in this respect, making Jupiter and its inner moons forbidden territory for human exploration for decades to come.

Among the more surprising of the many Voyager discoveries at Jupiter was a ring—a thin disc-shaped swarm of tiny particles—surrounding the planet. Made up mostly of dustlike bits of matter that become visible only when illuminated from behind by rays of sunlight, the ring was accidentally photographed as Voyager passed Jupiter's night side.

43

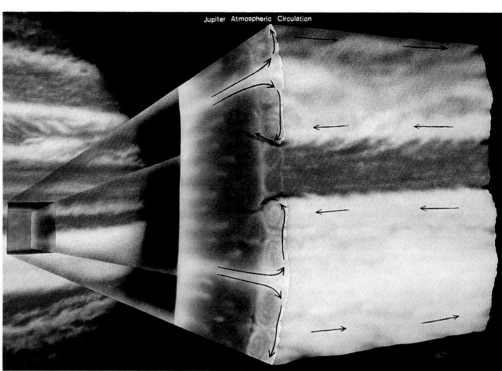

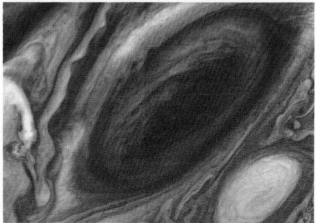

Nowhere in the solar system is the cloudscape as active or as colourful as on Jupiter, where booming winds and powerful vertical drafts whip the ammonia and ammonium-hydro-sulphide clouds into an ocean of unearthly beauty, **far left***. The clear hydrogen air allows sight lines hundreds of miles long on this largest of the planets. Above it all, Jupiter's moons Io and Europa are seen as dual crescents in the sapphire sky. The major atmospheric feature on Jupiter, known as the Great Red Spot,* **centre***, is a whirlpool-like storm centre larger than Earth. Jupiter's rapid rotation spins the clouds into distinct belts,* **above** *and* **left***, that can be seen from Earth with a 50-power telescope.*

45

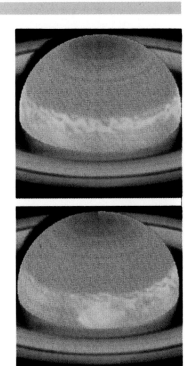

In November 1990, just after the Hubble Space Telescope began returning scientific data to Earth, Saturn's atmosphere became more active than it had been in 30 years. This pair of Hubble images, **above,** *revealing far more detail than Earth-based telescopes, shows half of one rotation of the planet. Near centre in the bottom image is a large white spot that initiated the outburst. It disappeared a few months later.* **Facing page:** *Human exploration of Saturn's satellite Mimas may not occur for a century, but there is no fundamental reason why it could not happen. Artist David Egge has anticipated the event by adding a few 21st-century adventurers to this scene of the icy moon's landscape.*

The structure is too faint to be seen from Earth, although once astronomers knew what to look for, electronically enhanced telescope imaging systems were able to detect it. The ring extends all the way down to the planet. Over time, pressure from sunlight and interaction with Jupiter's magnetic field cause the ring particles to spiral slowly inward and ultimately settle in Jupiter's upper atmosphere. There must be a source to replace the lost particles; otherwise, the ring would have disappeared after a few million years. The storehouse could be the inner small moons of Jupiter, whose surfaces are gradually worn down through constant contact with orbiting debris in their vicinity, or Jupiter's moon Io, which has its own amazing way of expelling material into space. (The moons of the giant planets are described in detail in Chapter 4.)

Saturn: Lord of the Rings

Saturn, the second largest planet in our solar system, is almost identical to Jupiter in general structure but somewhat cooler and less active in its interior. Because it is more remote from the warming influence of the sun, a high-altitude haze layer (cirrus-type clouds of ammonia-ice crystals) much thicker than Jupiter's veils the planet's face, producing a far blander appearance. The reduced interior heat generates less turbulent surface eddies, so the overall effect is a quieter atmospheric circulation. Although less impressive than Jupiter in size and surface features, Saturn is endowed with a hauntingly beautiful ornamentation—its rings—one of the universe's most dazzling creations.

Saturn's rings are composed of swarms of icy particles ranging from tiny crystals like those in an ice fog to house-sized chunks. Each has its own orbit around Saturn, although a gentle jostling occurs as the particles are affected by the gravitational pull of Saturn's major moons, each other and various forces, some of which have not yet been satisfactorily explained (but apparently produce the thousands of divisions in the ring structure).

The rings are truly enormous in extent. From edge to edge, they span a distance equivalent to two-thirds of the gulf between Earth and the moon. Yet the particles in the rings seldom stray more than 150 feet from a perfectly flat disc, making the structure about as thick as the height of a 30-storey building. A scale model of the rings made of paper the thickness of one of the pages in this book would be larger than a football field. The ring particles are not spread in a random haze around Saturn because of the distribution of mass within Saturn itself. Saturn's rotation period is almost an hour longer than Jupiter's, but because Saturn is the least dense of the gaseous giants, its material is more easily displaced by the rotational forces. The result is that Saturn is even more compressed at the poles and bulged at the equator than Jupiter; Saturn's difference is 12 percent. Thus a body orbiting around Saturn "feels" a greater gravitational pull as it passes over the equatorial zone than it does over the polar regions because there is more material below it. The path of greatest orbital stability is a nearly circular one above the most massive sector of the planet, precisely at its equator.

Within the rings, collisions gradually grind down the larger particles. Meanwhile, the smallest particles tend to stick to one another and increase in size through accretion. An equilibrium between the two processes has resulted. The very largest particles are near the rings' outer edge, where accretion is more effective. But the rings are too dense and too disrupted by the gravitational influences of Saturn's satellites to permit the process to escalate into the birth of a major Saturnian moon.

An examination of the rings by a spacesuited astronaut outfitted with a propulsion backpack for manoeuvring would be one of the most exquisitely beautiful excursions that a human could take in exploring other worlds. A spaceship orbiting Saturn could bring a ring explorer within a few miles of the ring structure without danger, since the orbital period of the spacecraft in an approximately circular path around Saturn would be identical to the orbital periods of the ring particles. A collision with any ring material under these conditions would be merely a nudge.

Stepping untethered outside the spacecraft, the ring explorer—using brief bursts of backpack propulsion—begins a slow glide toward the glittering plain of ring material extending seemingly to infinity. Descending into the ring structure, the explorer touches down on a large ring particle and, with one smooth push of the foot, is propelled on to the next major piece. In this slow-motion ballet, the explorer would move through a blizzard of particles—not falling anywhere, but frozen in suspension—gently bouncing off the front of the spacesuit and faceplate. Simply floating in the rings' golden gravel and being carried with it around

*Girded by its magnificent rings, Saturn is one of nature's most exquisite spectacles. These Voyager images show the planet with three of its moons, **top**, one casting its shadow on Saturn, and, **bottom**, the delicate ring structure, as finely grooved as a phonograph record. Saturn is an enormous planet. From edge to edge, the full extent of the rings would span two-thirds of the distance from Earth to the moon.*

the solar system's second largest planet would be an intoxicating experience: the partly hidden sun glinting off the multitude of particles, the gravitational symphony of collective motion that carries them around the planet, the feeling of being surrounded yet freely suspended in space—all disguise the fact that everything is whirling around the planet at 45,000 miles per hour.

For every house-sized ring boulder, there are a million the size of a baseball and trillions the size of a grain of sand. In denser sections, the baseball-sized particles would be separated by just a few feet, while the house-sized ones would be relatively rare, sometimes miles apart. The ring material resembles snow and is largely water ice. If the entire ring structure could be melted and refrozen as a solid ring, it would be less than two feet thick.

Uranus: Aquamarine Giant

Although it dwarfs Earth, Uranus is so remote that it appears as merely a pale greenish ball even in the largest telescopes, its five major moons tiny dots scattered around it. Not even the planet's rotation period—its day—was known with any accuracy prior to the first spacecraft encounter by Voyager 2 in 1986. Voyager's cameras detected several inconspicuous clouds in the planet's cloak of aquamarine "smog." Monitoring these clouds for a few rotations revealed that the planet turns in 16.7 hours. Nothing like the colourful cloud bands of Jupiter and Saturn were seen by Voyager. The lack of cloud activity can be partly attributed to Uranus's weak internal heat engine. The other factor is a haze of methane-ice crystals, a sort of ice fog, that further acts to block what lies beneath. Some rare towering clouds, elevated by exceptionally energetic eddies from below, were visible to Voyager —and these were seen only after substantial computer processing yielded details below the discrimination threshold of human vision.

By earthly standards, Uranus is mammoth. It is 63 times larger than Earth in volume, 14 times its mass and 4 times its diameter. A rocky structure somewhat bigger than Earth forms Uranus's core, which is encased in a much greater mass of icy slush— mostly water mixed with liquid methane, carbon monoxide and ammonia—thousands of miles thick. Overlying that is an equally thick, soupy atmosphere seven-eighths hydrogen, one-eighth helium, laced with methane, ammonia, ethane, acet-

ylene and ethylene. It is the smoggy, generally featureless top layer of that greenish atmosphere that the Voyager photographs displayed.

If astronauts of some future century ever explore Uranus, they will first encounter the methane haze as they descend into the atmosphere. It will look similar to the smog that sometimes permeates the air of Los Angeles as seen from an airplane window. Fifty miles below that is a methane-ice-crystal fog. The temperature here is minus 354 degrees F, only 95 degrees F above absolute zero. At an as-yet-undetermined level farther down, the methane cloud deck begins where the temperature rises and the atmospheric pressure is comparable with that at the Earth's surface. The methane below that

A celestial sea of moonlets, some as small as dust motes, whirl around Saturn in a thin plane that forms its spectacular rings. At this viewing angle, the sun illuminates clouds of smaller particles lifted from the ring plane by electrical forces similar to static electricity. In some of the Voyager images of Saturn, these clouds are seen as "spokes" in the rings. Ring particles are largely water ice, making them excellent reflectors of sunlight. Particles closest to Saturn orbit the planet in 5.6 hours; those at the rings' outer edge complete one circuit in 14.2 hours.

49

is in a liquid state, perhaps as a mist, with water-ice-crystal clouds. Deeper still, the water is liquid as well, probably as droplets suspended in the hydrogen and helium atmosphere. If the explorers continue down into the total blackness, they will find that the atmosphere increases in density, then turns into a slushy ocean of water, methane, acetylene, ammonia and probably liquid carbon monoxide that extends through to the solid core. But as on Jupiter and Saturn, the surface could never be reached. Well above it, both pressure and temperature rise to levels that would crush like tinfoil even a heavily armoured submersible.

Among the six planets that have been visited by spacecraft, only Uranus has a weirdly tipped rotation axis, which is just eight degrees from being horizontal. The flopped axis could have been the result of a collision with a body the size of Earth during the primordial era. The planet seems to roll around the sun in its 84-year orbit, rather than spinning in a near-vertical position as Earth does. For stretches of time lasting up to 42 years, parts of one hemisphere are completely out of sunlight. During the 1980s and 1990s, the planet's south pole is pointed almost directly at the sun. However, the amount of heat received from the sun is minuscule. To an explorer near Uranus, the sun would appear as a tiny disc the size of a pinhead held at arm's length. Sunlight, at one-quarter of 1 percent of its intensity at Earth, would illuminate the scene to a level similar to the gloom during a thunderstorm.

Uranus's system of nine skinny rings is but a shadow of Saturn's magnificent necklace. Virtually invisible from Earth, the Uranian rings were discovered by accident in 1977 when Uranus passed in front of a star and the star's light was dimmed by the previously unknown structure. The rings are narrow and as black as soot, the opposite of Saturn's. Another difference from Saturn's rings (and Jupiter's, which are composed entirely of fine dust) is the particle size. When Voyager's radio signal was deliberately passed through the rings, analyses of the resulting distortion indicated that they consist of refrigerator-sized and larger boulders of (presumably) the same icy mix as the satellites.

Neptune: Last of the Giants

Five months before Voyager 2's encounter with Neptune in August 1989, the spacecraft's cameras picked up a dark patch bigger than Asia floating in

Neptune's blue atmosphere. Neptune is so remote that no specific atmospheric feature had ever been seen from Earth. But the Great Dark Spot, as the huge feature is now called, proved to be an atmospheric eddy like Jupiter's Great Red Spot. Remarkably, the Dark Spot has exactly the same proportions as Jupiter's Red Spot and lies at the same latitude south of the equator. Just as Jupiter's spot is a dark shade of the creamy Jovian clouds, the Neptune spot is a dark shade of that planet's blue methane-rich atmosphere. Evidently, if the churning atmosphere of a gas giant is going to make a big spot, there is a natural size and position for it.

Methane clouds and haze in Neptune's upper atmosphere absorb red light and reflect blue light. Sunlight contains both colours, so an observer looking down on Neptune sees only the blue reflected light. Uranus's aquamarine hue is also caused by methane haze, but with a slightly different mix of other compounds. Methane (natural gas) liquefies at temperatures around minus 280 degrees F and solidifies a few degrees below that. Neptune's upper atmosphere is near this temperature and shrouds the planet in a blue methane mist. A few high cirrus-type clouds of white methane crystals stream above the main blue deck.

Five rings circle Neptune well above the upper atmosphere, three thread-thin ones and two so diffuse that they could barely be detected by Voyager 2. Rings had been suspected, both because stars had flashed off as they disappeared behind them (as seen from Earth) and because all the other giant planets are decorated by ring systems. Clearly, gas giant planets and rings are a normal arrangement.

Below the visible surface of Neptune is a bottomless atmosphere of hydrogen and helium that, thousands of miles down, eventually merges with a hot slush of hydrogen, helium and water ice sloshing around inside the planet. This core region rotates once in 16 hours, but the outer atmosphere spins more slowly and at different rates at various latitudes. The differential rotation rates create friction, which produces heat inside the planet. The heat percolates to the surface and generates turbulence —weather—that Voyager saw as the Great Dark Spot and several smaller eddies and bands.

The marked differences between Neptune's churning atmosphere and featureless Uranus can be partly explained by an aerosol haze on Uranus that is not present on Neptune. However, the main

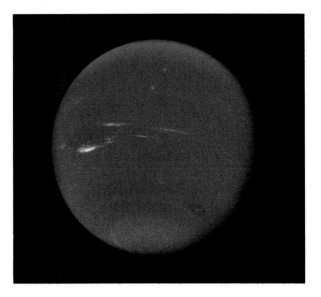

difference is that heat reaching the surface from within Neptune exceeds the heat received from the sun, whereas Uranus seems to be the reverse. For comparison, Jupiter has about 50 times Neptune's internal-energy outflow as well as receiving 20 times as much solar radiation, hence its more active cloud structure. Weak as these sources are on Neptune, there is no other way that the features we see on the blue giant could be produced.

Pluto: Enigmatic Mini-Planet

Cruising the outer rim of the known planetary system, Pluto is an enigma. It represents a third class of planet, different from the terrestrial worlds and from the gas giants. There is even some question as to whether it should be classified as a planet at all. It is far smaller than any of the others and smaller than seven planetary satellites, being only two-thirds the diameter of the Earth's moon. Most planetary astronomers define a planet as a body exceeding 600 miles in diameter (larger than the biggest asteroid) with its own orbit about the sun. Pluto barely qualifies. And its orbit is peculiar. It has by far the most elliptical path of any planet in the solar system, actually sweeping inside the orbit of Neptune at times. Between 1979 and 1999, Pluto is the eighth planet from the sun, Neptune the ninth.

Considering how insignificant Pluto is compared with the gas giants whose realm it shares—or compared with *any* planet in the solar system—I think it should be regarded as more closely related to comets than planets. I will explain why in Chapter 4.

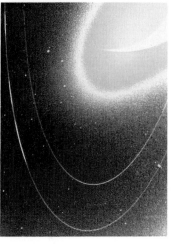

Slicing the blackness of space like a huge scimitar, the crescent of Uranus, **facing page**, *loomed before the cameras of Voyager 2 just after the spacecraft passed the planet in January 1986. Neptune,* **above left**, *seen by Voyager 2 in August 1989, proved to have a far more active atmosphere, which includes the Asia-sized Great Dark Spot. Cirruslike clouds above Neptune's ocean of blue methane mist were seen in detail,* **top**, *just before Voyager dashed by the planet. Neptune's uneven, skinny rings,* **bottom**, *are visible below the overexposed image of Neptune in another Voyager portrait of the eighth planet.*

51

ICE WORLDS

An alien spaceship approaches the solar system. A huge projection device in the ship's flight-deck control room displays the sun and surrounding space. Although still well beyond Pluto's orbit, the ship's powerful sensors detected Jupiter long ago and are now monitoring Saturn, Uranus and Neptune. Images of the giant planets flash into view—crescent Jupiter; Saturn's huge shadow falling across its own rings; Uranus and its moons—along with a dizzying array of cryptic statistical information and analyses. The astronomer on duty turns to his co-workers and casually announces: "Single yellow main-sequence star with one large primordial gas giant and three smaller ones . . . elsewhere, there is some debris that we can examine when we're closer in."

Some debris! That's us. Earth.

Yet Earthlings frequently assess their own environment in precisely the same manner. There is more to the solar system than the sun and the nine planets. Some of the satellites of Jupiter and Saturn are as large as the planet Mercury and, along with the rest of the moons of the giant planets, form miniature "solar" systems of striking diversity. There are oceans, volcanoes, geysers and craters in abundance. These are worlds that bring far more variety to the solar system than the planets alone.

The satellite systems of Jupiter, Saturn and Uranus bear more than a superficial resemblance to a scaled-down solar system: they probably formed in a similar way. The contracting clouds of material around the emerging planets made them pseudo-stars for a time, each with its own miniature solar nebula. In a further parallel, the billiard-table flatness of the planetary orbits is cloned in the satellite families of Jupiter, Saturn and Uranus.

Jupiter's satellite system fits into two distinct categories: very large and very small, with nothing in between. The four big moons, discovered by Galileo in 1610 shortly after the invention of the telescope, range from the size of the Earth's moon to the dimensions of Mercury. These Galilean satellites are worlds unto themselves.

Callisto: Battered Iceball

Outermost of the Galilean moons and the only one far enough from Jupiter's intense radiation belts to be a possible landing site for Earth explorers unprotected by cumbersome shielding, Callisto resembles the Earth's moon or Mercury, but it is more cratered. Craters are jammed rampart-to-rampart everywhere on Callisto, in greater profusion than on any other body in the solar system.

A walk on Callisto would, in many respects, be like an exploration of the Earth's moon: the two bodies have a similar surface gravity, and neither has an atmosphere. But the similarity would end upon examination of a collection of surface materials. Bits of rubble and chunks of rock gathered from Callisto would melt when exposed to room temperature. Callisto's "rock" is mostly water ice with lunar-like dirt mixed in. At Callisto's surface temperature of minus 230 degrees F, water ice acquires the stiffness of rock, rather than the more plastic characteristics of glaciers on Earth. Deep in Callisto's interior, a silicate rocky body probably exists, but the bulk of the planet beyond the core is water ice.

. . . worlds unthought of until the searching mind of science laid them open to mankind.

WILLIAM WORDSWORTH

One of Io's volcanoes spews boiling sulphur and sulphur dioxide into a spectacular 100-mile-high fountain.

53

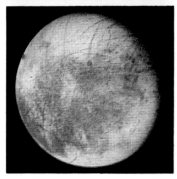

Three of Jupiter's four major satellites, **top to bottom:** *Callisto, Ganymede and Europa.*

The crater-scarred surface of Callisto appears much as it must have looked four billion years ago. There is less evidence of change on its surface than there is on the Earth's moon. If the other three major Jovian satellites were like Callisto, Jupiter's system would be interesting but not too much different from astronomers' expectations. The two Voyager spacecraft revealed that the four moons are not all the same and also that their surfaces represent some of the most remarkable and unexpected finds in the history of astronomy.

Ganymede: Jupiter's Giant Moon

Ganymede, the next Galilean satellite inward from Callisto, is the largest moon in the solar system, slightly larger than the planet Mercury. Like Callisto, it is a cratered world with a mainly water-ice surface darkened by dirt. Some of the major crater impacts punched through the light brown soil-ice composite, exposing brighter, cleaner ice underneath. The ejected icy material is splattered around these craters in what astronomers call ejecta blankets. But there is more than craters on Ganymede. Large areas of ribbed and ridged icy material are spread like continents over the satellite. These regions have fewer craters than other sectors and are clearly younger. Planetologists suggest that internal heat from Ganymede produced the features, which look like glacial flows on Earth.

Ganymede and Callisto are both believed to be about half water ice and half carbonaceous (carbon-rich) soil. Originally, these worlds may have been fairly homogeneous internally, but radioactive materials in the soil released heat soon after the moons' formation, partly melting them and allowing the heavier matter to sink to the core and the ice to become the predominant surface material. This situation remains today. Perhaps due to its slightly greater mass, Ganymede retained enough heat for a second wave of melting to form the grooved terrain. The transition zone between the icy surface layers and the soil cores deep within the moons may be a slushy mixture of ice and dirt.

Europa: Ice-Encased Ocean World

Europa is the smallest of the Galilean moons and the second out from Jupiter. Until 1979, it was just another astronomy-textbook statistic. Then came the Voyager 2 close-up images, and within days, Eu-

ropa was transformed—in our perception at least—into one of the solar system's most intriguing worlds. The biggest initial surprise was the almost total lack of detail. From a distance, Europa looks like a white cue ball. At close range, the only visible features are thin, kinked, brown lines resembling cracks in an eggshell. This analogy is not far off the mark.

The surface of Europa is almost pure water ice, but the absence of craters (only four small craters were found in the Voyager photographs) indicates the ice is far less rigid than that of Ganymede or Callisto and more closely resembles the Antarctic icecap. The eggshell analogy could well be completely accurate, since the ice is probably only 50 miles thick—a true shell around what is likely a subsurface ocean of liquid water that in turn encases a carbonaceous and rocky core. The interior of Europa appears to be significantly warmer than Ganymede's or Callisto's due to a process called tidal heating. Gravitational interplay among the Galilean satellites causes the moons to vary slightly from perfectly circular orbits around Jupiter. Europa's orbit does not deviate much from being circular, but Jupiter's gravity is extremely strong and stretches Europa out of shape—less so when it is at its farthest distance from the planet. These tidal forces on Europa generate interior heat, which melts the ice almost to the surface. The marks on Europa's icy face may in fact be cracks through which melted material oozes from below.

A few months after Voyager 2's encounter with Jupiter in 1979, when the best images of Europa were obtained, science writer Richard Hoagland advanced the startling idea that Europa's subsurface ocean could harbour life. Hoagland suggested that "Europa wasn't always encased in ice. Near the origin of the solar system, Jupiter was more like a miniature sun than a planet, shedding enough heat that, in combination with the processes occurring on Europa, would have allowed [Europa's] surface to be covered in an ocean as the Earth was early in its history." Jupiter's early heat was produced by the compression of the material forming the giant planet. Just as the primal sun was far more radiant than it is today, so the heat generated by Jupiter is but a shadow of its former intensity.

During this warm phase, some 4.6 billion years ago, life processes may have gained a toehold on Europa, as they presumably did on Earth. According to Hoagland's scenario, "the ocean froze, but not before the primordial 'soup' was locked in, and

Jupiter's intriguing satellite Europa will be examined with high-resolution electronic cameras and other instrumentation on the Galileo spacecraft, **above left**, scheduled to begin orbiting Jupiter in the mid-1990s. One theory that emerged in the wake of the Voyager explorations of 1979 suggests that for a few million years after Jupiter's formation, **above**, the giant planet radiated enough heat that Europa remained water-covered, perhaps offering a chance for life forms to gain a foothold in the outer solar system.

it has been there—possibly evolving biologically—ever since.'' Hoagland's concept of life in the Europan ocean was initially dismissed by planetologists, but since then, several have come forward with their own versions of the idea. Science fiction superstar Arthur C. Clarke was so intrigued by Hoagland's theory that life within Europa became a central feature of his novel *2010: Odyssey Two*.

Today, the theory stands as just that, a theory, but it is bolstered by the discovery (which coincidentally occurred around the same time that the Voyagers passed Jupiter) of a form of life on the Earth's ocean floor that exists in total blackness, sustained entirely by chemical, rather than solar, energy. These creatures live around warm-water vents in deep-sea midocean rifts, relying on the sulphur and oxygen in the mineral-rich water for the energy required to support them. The seawater seeps through porous rock on the ocean floor, becomes heated and saturated with minerals from exposure

to the hot rock below, then fountains up through chimneylike vents. As the hot water meets cold seawater, the minerals precipitate onto the surroundings, producing a fertile crucible for the chemically based organisms. Until this discovery, sunlight was thought to be the sole energy source for all higher life forms on Earth.

On Europa, the internal heat that keeps the ice-topped ocean from freezing could create the same type of ocean-bottom vents. If life began in the primal open oceans that might have existed when Jupiter was a miniature sun, the organisms could have adapted to obtaining sustenance at the seafloor by chemical means after the surface froze. One tantalizing footnote to these speculations was provided by Voyager 2's final image of Europa, which shows what could be a vapour cloud from a vent in the ice. The Galileo Orbiter is designed to gather more images of Europa during its circuits of Jupiter in the 1990s.

In 1979, Voyager 1 captured a stunning view of Jupiter's cloudscape acting as a back-drop for the planet's moons Io (orange) and Europa (white), **right**. Few astronomical images rival this one in expressing the beauty of nature. A close-up view of Io, **bottom left**, also taken by Voyager 1, reveals a tortured landscape of volca-noes, rivers of molten sulphur and white beds of solid sulphur deposits. Io's huge volcano Pele, **bottom right**, ejects 10 tons of molten sulphur and sulphur dioxide each second. Part of its plume, 200 miles above the volcano, can be seen against the black sky.

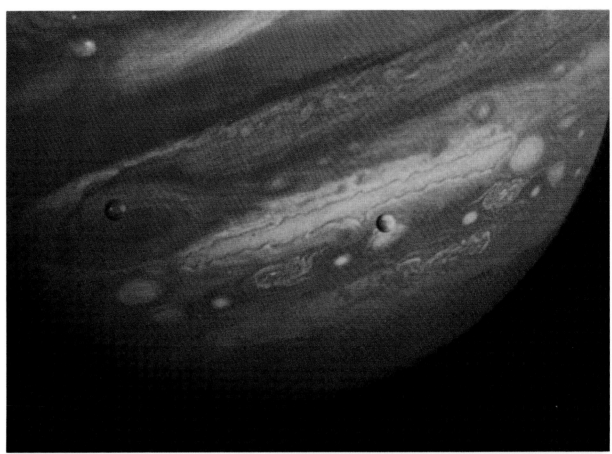

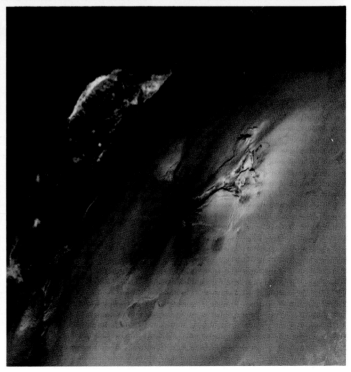

Io: Focus of Furious Forces

If Jupiter's satellite Io (EYE-oh) had ever been described in a science fiction novel, it would have been dismissed as too bizarre to be real. Here is a world where a dozen giant volcanoes are simultaneously erupting with such violence that material is ejected more than 100 miles into the sky. Molten sulphur oozes down the flanks of the volcanoes and spreads into the valleys below. So much matter is being expelled by the volcanic forces that this moon has literally turned itself inside out two or three times during its existence. Over the surface of Io, radiation from surrounding space is so intense that lead shielding a foot thick would be barely adequate protection for a human.

Io is about the size of our moon and is the nearest of the Galilean satellites to Jupiter. It was a virtually unknown object until Voyager 1 sped to within 12,000 miles of it on March 5, 1979. The discovery of the real Io ranks as one of the major triumphs of modern space exploration. In less than 10 days, astronomers went from thinking Io was similar to the Earth's moon to the realization that it is totally different. As a reporter at the Jet Propulsion Laboratory in Pasadena, California—Mission Control for the Voyager probes—I witnessed the drama unfold firsthand. Voyager hurtled closer and closer to Io in early March, its electronic cameras steadily gaining resolution. But the images revealed no craters. (The more craters, the longer the object has been battered and therefore the older its sur-

face.) The surface was a colourful quilt of white, brown and orange splotches that one Voyager scientist described as a "celestial pizza."

Three days after Voyager's closest approach to Io, Linda Morabito and Steve Synnott of the Voyager optical navigation team were comparing the position of Io with background stars, a procedure necessary for refining knowledge of the spacecraft's position and the pointing accuracy of its cameras. Using computer enhancement, they overexposed the images to bring up the background stars. Suddenly, an umbrella-shaped plume hovering above Io's surface materialized. Morabito identified it as an eruptive cloud from a volcano. Other possibilities, such as a glitch in the imaging system, were quickly ruled out, and the announcement was made: we were gazing down on a live volcano, the first ever seen on another world.

Ultimately, 12 active volcanic centres were discovered on Io, spewing enough material to blanket this tortured moon at the rate of several inches per century. Just as Europa's interior is heated by the tidal forces caused by Jupiter, so Io is affected. But the tides raised on Io are far greater, enough to melt its interior completely and to trigger the global scourge of volcanoes.

Io is significantly denser than the other Galilean moons because long ago it turned its ice into steam and blasted it into space through volcanic activity. Apparently, only sulphurous rocky material remains. A molten interior of sulphur and rocky lavas is overlaid by a thin crust through which the volcanoes eject their contents. The colourful surface of the moon is due to the various hues that sulphur takes on at different temperatures. At more than 1,100 degrees F, sulphur is black, which is usually the colour of the volcanic vent regions. As it cools, sulphur turns lighter, first brown and then bright orange, and these shades correspond with the colours of the flanks of the volcanic vents. At room temperature, sulphur is yellow but becomes white in the extreme cold at Io's distance from the sun. This is the colour of the inactive plains of Io. Like exploration of the surface of Venus, an expedition to Io by humans is probably a long way in the future.

Satellite Inventory

The remaining 12 Jovian moons—debris compared with the Galilean satellites—are from 125 to 10 miles in diameter. The little moons can be sub-

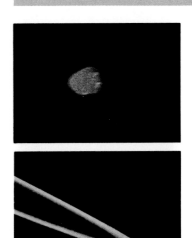

*A dormant volcanic caldera on Io, **left**, is caked in sulphur deposits, the chief component of the satellite's fierce eruptions. Enough sulphur is belched out of Io's volanoes to have long ago coated its small neighbour moon, Amalthea, **top**, a 165-mile-long potato-shaped body first seen from Earth in 1892. This view is the best shot of the small moon taken by Voyager 1 during its 1979 Jupiter flyby. Amalthea orbits Jupiter in just under 12 hours, keeping its most pointed end always aimed at the big planet. Voyager was first to see Jupiter's ring, **bottom**, shown prominently here but, in reality, a very inconspicuous feature.*

57

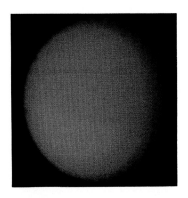

*Cloaked in mist, smog and clouds, Saturn's moon Titan, **above**, is the only known celestial object with an atmosphere approximately the same density and surface pressure as the Earth's, and its icy surface may have lakes of liquid methane and ethane, **left**. The illustration depicts the descent of the proposed Cassini probe of Titan's atmosphere and surface in the 1990s. The probe is releasing a parachute to allow a slow descent to the landscape below. No other Saturn moon has an atmosphere, but Enceladus, **far left**, shows evidence of surface modification—a smooth plain almost devoid of craters—perhaps caused by steam and water geysers re-coating the surface.*

divided into three groups, four inside the orbit of Io, four clustered at six times Callisto's distance from Jupiter and four at twice that distance.

Saturn's family is more diverse, with one moon the size of the planet Mercury, four medium-sized ones, from one-quarter to one-half the diameter of the Earth's moon, and a dozen smaller than that. Just as at Jupiter, the Voyager spacecraft provided stunning information about Saturn's satellites. The transition from the unknown to the known was even more dramatic because for centuries, Saturn's moons, at twice Jupiter's distance, were displayed to earthbound telescopes as mere shimmering points of light. According to what we now know, Titan—Saturn's largest satellite—is by far the most intriguing of the group.

Titan: Glaciers and Methane Rain

In 1944, American astronomer Gerard Kuiper detected methane gas around Titan. That, plus the fact that Titan is about the size of the planet Mercury and orbits Saturn once in 16 days, was basically all that was known prior to Voyagers 1 and 2. Now, it ranks as one of the most favoured sites in the solar system for further exploration and is in many ways more interesting than Saturn itself.

Titan is a featureless globe, a world cloaked in a yellowish brown smog of upper-atmosphere aerosols and hydrocarbons. Voyager experiments were able to penetrate the veil to reveal that Titan is the only place in the solar system which has an atmosphere even remotely similar in composition and density to the Earth's. Our atmosphere is 78 percent nitrogen and 21 percent oxygen. Titan's atmosphere is about 95 percent nitrogen and 5 percent methane, with traces of carbon compounds. The surface atmospheric pressure on Titan is 1½ times what it is on our planet. The big difference is temperature. The Earth's average surface temperature is just above the freezing point of water, while Titan's, at minus 292 degrees F, is barely above the freezing point of methane. As water on Earth exists in solid, liquid and gaseous states, so methane on Titan is solid, liquid and gas. There must be glaciers of methane ice, oceans of liquid

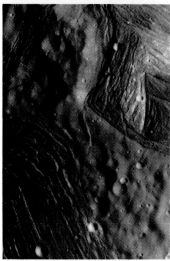

*Images of two distant ice worlds—Saturn's satellite Dione, **top**, and Uranus's Miranda—reveal landscapes totally unknown prior to the development of sophisticated robot spacecraft. The image of Miranda shows craters and hillocks the size of major sport stadiums. Both moons are largely ice, but at the frigid temperatures in the outer solar system, ice behaves like rock, rigidly sustaining craters billions of years old.*

methane, and methane vapour in the lower levels of the nitrogen atmosphere that could fall as methane rain. Methane—the stuff we know as natural gas—serves the same role on Titan as water does on Earth. Ethane, a compound chemically related to methane, is likely present in large quantities, perhaps forming 5 to 10 percent of Titan's oceans.

Humans protected by well-insulated, heated spacesuits would have no problem exploring Titan. The ground would be an amalgam of methane ice and water ice that, on casual inspection, would seem identical to snow. When it snows on Titan, the flakes would hardly seem to be falling, since gravity is only one-sixth the Earth's, about the same as on the surface of our moon. The low-gravity and dense-atmosphere combination makes Titan the best place in the solar system for flying. Aircraft with small motors could easily stay aloft, and gliders would soar almost endlessly. Human-powered flight might even be possible. There are some hindrances, however, such as the need for floodlights, flashlights and helmet lamps to penetrate the gloom. By the time light from the sun filters through Titan's dense smog and clouds, it illuminates the surface with a glow barely brighter than our full moon's.

Titan probably has an extremely stable weather system due to its vast distance from the sun and the thickness and insulating properties of the atmosphere and clouds. The sky would appear a dull brownish colour, likely featureless, without any definite indication of clouds, similar to a drab winter day on Earth but less than 1 percent as bright. An expedition on the surface of Titan would resemble an exploration of Antarctica: intense cold, seemingly endless stretches of ice and snow, relatively little change in weather. And on Titan, a day is 16 Earth days long—8 days of light and 8 days of pitch-darkness. There are probably ponds or lakes of liquid methane. One interesting property of a methane lake is that it never freezes over: methane ice sinks, so the ice builds up from the bottom. Boats may be a 21st-century mode of transport on Titan.

Could life in any conceivable form exist in Titan's supercold environment? Prolific science fiction/science author Isaac Asimov proposed how it could happen in a remarkable speculative article in 1961, although he did not have Titan in mind. He realized that liquid methane will not dissolve the same kind of materials that water does, so biochemistry on Titan could not be based on proteins and nucleic acids. But methane will dissolve lipids, a class of compounds including oils and fats. On Earth, lipids form molecules whose complexity is comparable to that of protein. Compounds that develop in Titan's methane environment would remain stable because there is no free oxygen in its atmosphere to break them down. Hence, a methane biology, wrote Asimov, might produce complex organisms that would clearly be a form of life.

Whether or not that has occurred, some of the atmospheric methane on Titan is likely being processed by chemical reactions induced by solar ultraviolet energy. The resulting hydrocarbons would rain down on the surface, leaving a layer of life precursors, perhaps as a brownish goo. One biologist has proposed that the material might accumulate in bogs up to 40 feet deep. Current theories suggest that primitive Earth, like Titan, had an oxygen-free atmosphere with significant amounts of methane, but the parallel is not complete, because the temperature on Earth was always higher. However, Titan may be the closest equivalent scientists have found to the conditions that provided the crucible for the emergence of life on Earth.

Titan's nitrogen atmosphere was as startling a discovery as the volcanoes of Io, but in retrospect, it makes sense. Just after Titan formed, it was probably much warmer than it is now, a time when there were liquid-ammonia oceans and gaseous ammonia was a major constituent of the atmosphere. To produce the nitrogen, the ammonia was either frozen to the ground or broken apart by solar ultraviolet radiation into hydrogen and nitrogen. The nitrogen would collect in the atmosphere, and the hydrogen would be lost into space. If it was ever warm enough in the early days of Titan to allow water to be liquid, life-precursor molecules could have formed in abundance. Whatever was produced must still be there, frozen in the water-ice continents.

Uranus's Satellites

Uranus's five major satellites—Miranda, Ariel, Umbriel, Titania and Oberon—all came under Voyager 2's scrutiny in 1986. Titania and Oberon are half the diameter of the Earth's moon; Umbriel and Ariel, about one-third as big; and Miranda, only one-seventh our moon's diameter. No details had ever been seen on any of these moons from telescopes on Earth, but Voyager's high-resolution electronic camera displayed Oberon, Titania and Umbriel in detail comparable to a binocular view

The utterly alien landscape of Triton, Neptune's largest moon, was captured in exquisite detail, **left**, by Voyager 2 during the spacecraft's encounter with Neptune in August 1989. At 30 times the Earth's distance from the sun, the temperature is only 65 degrees F above absolute zero—even colder than Pluto, which is cruising inside Neptune's orbit until 1999. Even at that temperature, there are seasons. It is spring in Triton's southern hemisphere (south is at right in this image), where an extensive cap of frozen nitrogen is slowly melting. The black streaks in the polar-cap area are downwind stains from active nitrogen/methane volcanoes. High-resolution view of Triton, **above**, shows a small area near the equator.

of our moon from Earth. All three have abundant craters like those on the Earth's moon.

Voyager swung close enough to Ariel to show features about the size of a small city. At that distance, some unexpected trenches and structures indicating glacierlike flows appeared. But it was Miranda that yielded the most surprises. Planetologists expected it to look like Saturn's moon Mimas, a crater-covered world the same size as Miranda located a similar distance from its parent planet. Instead, Voyager presented what one planetary geologist described as a bizarre hybrid of the geology of the planets Mercury and Mars and some of the moons of Jupiter and Saturn. Miranda sports canyons, ribbed flow features, crevices and sectors almost devoid of craters.

The Uranus satellite system seems to confirm, once again, the bull's-eye pattern of the solar nebula that dictated the composition of the planets and their satellites. The inner planets and the Earth's moon are rocky, but Jupiter's large moons, though similar in size, have mixtures of rock and water ice. Continuing outward, the moons of Saturn, including Titan, have less rock and more ice. These differences reflect the cooler conditions that existed farther from the sun when these bodies formed. At Uranus, the temperature was low enough that methane, carbon monoxide and ammonia were available in addition to water as solid moon- and planet-building materials. Water ice is as hard as stone in the frigid realm near Uranus. But methane, ammonia and carbon-monoxide ice could flow in some way analogous to glacial movement on Earth, which may explain some of the strange terrain on Miranda and Ariel.

Triton: Neptune's Companion

The triumph of Voyager 2's 12-year odyssey from Earth to Jupiter to Saturn to Uranus and, finally, to Neptune in 1989 is that the remarkable spacecraft forever transformed worlds in the remote outer

Methane ice on the surface of Pluto is slowly vaporized by the feeble radiation from the remote sun, which has the same apparent brightness as a high-intensity streetlight. The temperature here is minus 370 degrees F—a comparative heat wave on Pluto, since it is at its nearest to the sun (30 times the Earth's distance from the sun) during the 1980s. In 125 years, Pluto will swing to 49 times the Earth-sun distance, and the temperature will be minus 410 degrees F, only 50 degrees above absolute zero. In many respects, Pluto is more closely related to comets than to the other eight planets of the solar system.

solar system from mere dots, as seen in our largest Earth-based telescopes, to real places, with volcanoes, canyons and icy plains.

The last of these transformations occurred when Voyager hurtled past Neptune's largest satellite, Triton, on August 25, 1989. Two hundred close-up pictures of Triton, transmitted to Earth from Voyager's on-board image-recording system, revealed an incredibly detailed portrait of an amazing world —the coldest place in the solar system—where water ice is as hard as granite and liquid nitrogen acts like water, forming a subsurface layer analogous to the water table beneath the Earth's surface. Natural heat from inside Triton, generated by tidal action from nearby Neptune and the decay of radioactive elements, is enough to sustain the liquid state of the subsurface nitrogen "groundwater."

Much of Triton is covered with brilliant blue-white nitrogen snow deposited from geyserlike eruptions of liquid nitrogen, which escapes to the surface through vents or cracks. Frozen methane (solid natural gas) is also released in these eruptions, leaving purple and black stains beside the vents. Dozens of the stains are visible in the Voyager photographs. They are elongated because of prevailing winds on Triton. Winds are not common on planetary satellites. Most, like our moon, are airless, but Triton sports a thin atmosphere of nitrogen and methane. The atmosphere is clear except for a few wispy clouds.

Unlike most other moons in the solar system, Triton has virtually no craters. Those that must have covered Triton billions of years ago have been erased by flooding from liquid-water "lava" perhaps two billion years ago and by more recent coverings of liquid methane and nitrogen.

Neptune may be off limits for human exploration, but Triton can be explored. Perhaps a century from now, astronauts will clamber over what scientists call the cantaloupe terrain or examine the vents thought to release melted nitrogen ice periodically, warmed by pressure and change of season. Even here, 30 times the Earth's distance from the sun, summer, however feeble, still arrives.

Future explorers will confirm or refute the idea (supported by the Voyager discoveries) that Triton was once an independent planet of ice and rock plying its own orbit about the sun. The theory suggests that Triton strayed too close to Neptune and was captured by the gravity of the more massive planet about four billion years ago. For the next bil-

lion years, it swung in a highly elongated orbit, crashing into any moons Neptune originally had beyond the six known to be near the planet. Each time Triton passed close to Neptune, tides (caused by the big planet's gravity) would be strongest. When Triton reached the elliptical orbit's far end, the tides would be much weaker. The constant pulling and relaxation of Triton would have generated enough internal heat, astronomers calculate, to turn the moon molten. Two to three billion years ago, through natural orbital evolution, Triton's orbit became circular, the tides subsided, and the surface froze. This scenario explains most of what we see, but regions on Triton like the bizarre cantaloupe terrain remain puzzling.

Planet X, Pluto and Comets

A suspected tenth planet at the edge of the solar system, dubbed Planet X by astronomers who predicted its existence earlier in this century, is a myth. It does not exist. The original evidence offered for Planet X was that observations of the positions of Uranus and Neptune indicated the two planets were deviating slightly from where they should be according to orbital theory. These deviations, the Planet X proponents suggested, could be accounted for by the gravitational influence of a tenth planet, larger than Earth, somewhere beyond Pluto. However, the discrepancies were the result of measurement and instrumental errors in determining the position of the two outer planets combined with an inaccurate estimate of the mass of Neptune. Once the correct mass of Neptune was obtained by Voyager 2 in 1989, researchers rechecked the old positional measures and uncovered the other errors. The locations of Uranus and Neptune as they glide along their orbits now match predictions exactly.

The search for Planet X has many parallels with the quest for the "missing" planet between Mars and Jupiter two centuries ago. Thousands of asteroids roam where a single planet was expected in the Mars-Jupiter gap. Similarly, a belt of comets exists where the Planet X hunters sought a single large body. The first of these comets was picked up in 1992 during a deliberate search by David Jewitt and Jane Luu using the University of Hawaii's 88-inch telescope. It is a chunk of primordial ice—a giant comet—about 100 miles in diameter orbiting the sun at 1.5 times Neptune's distance. In 1993,

several more objects of similar size were detected at about the same distance. Known as the Kuiper belt, after Gerard Kuiper, the astronomer who predicted its existence in 1951, the zone undoubtedly contains thousands of comets greater than a few miles across.

In another of the many ironies in the history of astronomical discovery, we were looking at a member of the Kuiper belt all along—Pluto. Pluto's elliptical orbit carries it from 0.98 to 1.66 times Neptune's distance from the sun, the same zone where the newly discovered Kuiper belt comets roam. Comets are giant snowballs (water, methane, ammonia and other ices), which is the best estimate of Pluto's composition.

Halley's Comet, the comet we know the most about, offers some insight into the nature of these objects from the icy depths at the outskirts of the solar system. At its most distant point from the sun, Halley's Comet is just outside Neptune's orbit, but at its closest, it zooms in to Venus's distance. This long, sausage-shaped elliptical orbit is typical of comets that can be observed from Earth. When a comet nears the sun, its surface is vaporized by solar radiation and expelled into a cloud that is swept back into the classic comet tail by the pressure of sunlight and the force of the solar wind (electrically charged particles emitted from the sun). But it is the comets that we don't see which interest us here. These are the ones that never get close enough to the sun for the vaporization process to occur. These are the *normal* comets. Halley's Comet and all the other known comets are freaks—comets that were disturbed from their normal orbits somewhere beyond Neptune.

Halley's flourish through the Earth's skies every 76 years is the sign of a dying comet. It loses the equivalent of several feet from its surface during every visit to the sun's vicinity. It can last for only several thousand circuits of the sun until its store of ices is exhausted. Eventually, all that will remain is a clod of dirt and rock. But comets that stay out beyond Neptune are pristine, basically the same as they were billions of years ago. The only ones we have seen are Pluto and its smaller companions. There could be as many as 100 trillion of them larger than the size of a major sporting arena, reaching out in a diffuse cloud to one light-year or more. Triton was likely a Pluto-sized comet captured by Neptune during the early days of the solar

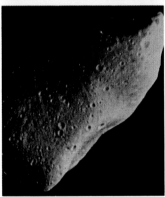

Asteroids are miniature planets that probably resemble Phobos, the small Martian moon, **top,** *which many astronomers suspect is an asteroid captured by the planet. The first close-up photograph of a real asteroid, Gaspra,* **above,** *was obtained by the Galileo space probe in 1991 on its way to Jupiter. The majority of the 2,000 known asteroids orbit in the zone between Mars and Jupiter, but some come within a few million miles of Earth. Phobos's irregular potato shape is typical of solar-system bodies under 150 miles in diameter, its gravity too weak to pull it into a spherical shape, as that of a more massive object (planet, star, large satellite) always does.*

system, when comets were more abundant. Indeed, Uranus and Neptune are themselves related to comets. Their compositions of water, methane, ammonia and hydrogen are in the same approximate ratio as those elements and compounds in the ices of comets.

During the solar system's formation, trillions and trillions of comets condensed from the solar nebula in the region where the outer planets formed. These were the planetesimals of the outer solar system, the building blocks of the giant planets. Once Uranus and Neptune were established, they became the gravitationally dominant bodies. Like tagteam wrestlers, they soon threw the rest of the comets out of the ring. Some were flung completely outside the solar system, while the rest were hurled from their nursery to form the Oort Cloud, beyond the influence of the planets. The Kuiper belt, just beyond Neptune, is believed to contain the comets that formed farther out and survived the mayhem nearer Neptune and Uranus.

The normal long-term motions of nearby stars, which occasionally bring them within two lightyears of the sun, combined with the gradual shifting of mass distribution in the arms of the Milky Way Galaxy around us occasionally provide enough gravitational deflection to disperse some of the comets from their remote outposts and dump them toward the inner solar system. That is how Halley's Comet and the other visible comets arrived.

Returning to Pluto, I must address the statement so often repeated in kids' astronomy books that "Pluto roams the outer reaches of the solar system in permanent darkness, the sun only a bright star in the sky." This is simply not true. The sun appears smaller than a pinhead held at arm's length, but it still sheds light equal to about 600 full moons. You would have no trouble reading this book by sunlight on a spaceport on Pluto. But it *is* cold, only 70 Fahrenheit degrees above absolute zero.

Pluto's 248-year orbit is so elliptical that its distance from the sun ranges from 2.8 billion to 4.6 billion miles, an effect that produces Pluto's seasons. The Plutonian summer occurs around perihelion, its closest point to the sun, when the methane and nitrogen ices on Pluto just reach the sublimation point—the transition from solid to gas—to form a thin atmosphere. Perihelion occurred on September 12, 1989, so Pluto's summer is now at its peak. To take advantage of this season of activity, NASA is planning a Pluto flyby early in the 21st century.

Asteroids

More than a million asteroids populate the zone between Mars and Jupiter. Ceres, the largest asteroid, is about the same width as France; the smallest are mere flying boulders. A few renegade asteroids roam inside Mars's orbit or beyond Jupiter, but the bulk of them have stable orbits in the Mars-Jupiter gap. When we look at a plan of the solar system, we see this as a real gap. Not only does it separate the giant planets from the smaller terrestrial worlds, but it's large enough to accommodate a planet easily. Yet none formed there. The prevailing theory is that Jupiter's gravitational influence so perturbed this region of the solar nebula that one large planet was never able to develop. Instead, a number of smaller ones did, some of which collided, and later the fragments collided, and so on, resulting in the range of sizes we see today.

The total mass of the asteroids is probably less than that of the moon. In composition, they range from chunks of nearly pure nickel/iron to carbonaceous bodies similar in some ways to garden dirt. Ceres is one of the original primordial mini-planets. There may be a few others, but most of the asteroids are fragments of those original bodies. In 1991, on its way to Jupiter, the Galileo spacecraft obtained the first close-up look at an asteroid (photograph at left). Gaspra is 12 miles long and 7 miles wide, approximately the same size as the two moons of Mars.

Of most interest to Earthlings are the several dozen asteroids that approach closer to Earth than either Mars or Venus. The largest, named Eros, is a brick-shaped chunk of asteroidal material three times as long as it is wide, approximately the dimensions of Manhattan Island. It cruises three times closer to Earth than Mars does. Smaller asteroids, less than a mile across, occasionally glide past at only a few times the moon's distance. These flying mountains are visible for only a few days as faint specks of light traversing the remote star fields. Searches turn up a couple of new Earth-approachers every year, indicating that there must be hundreds. All of them originate in the main asteroid belt. Nudged by a close encounter with a companion, they swing our way. Statistically, such an object should clobber Earth every million years or so, exploding on impact with a force far exceeding the power of the largest hydrogen bomb—reason enough to find out more about these renegades.

When Halley's Comet returned to the inner solar system in March 1986, it was greeted by five spacecraft that had been launched months earlier. Back on Earth, millions of stargazers watched as the comet slowly cruised through the constellations. Since the comet was three times farther away from Earth than it was on its previous visit in 1910, it was never a spectacle. This photograph, taken from a mountaintop in New Mexico, shows both the blue gas tail and the yellow dust tail typical of most comets.

COSMIC FURNACES

CHAPTER FIVE

There was a time when talking about other stars and planets could be dangerous to one's health. In 1600, Giordano Bruno was burned at the stake partly because he maintained that "countless suns" and "an infinity of worlds" exist, a heretical notion at the time.

Stars are suns, of course, and there are more of them in the universe than there are grains of sand on all the beaches on Earth. They are the citizens of the galaxies. Like people, some stars are young, some old, some big, some small, some more noticed than others. If a census of the Milky Way Galaxy's hundreds of billions of stars could be made, the sun would come out quite respectably in the top 5 percent—larger and brighter than average. It would be further distinguished because our sun does not have a companion star. At least two-thirds of the stars in our sector of the galaxy are multiple-star systems, usually binaries, but sometimes triple and quadruple groupings exist in a complex interweaving of gravitational forces. Rare, though by no means unknown, are families of five or six stars, all bound together by gravity's grip.

These multiple-star systems are not analogues of solar systems, however. They arrange themselves in a different way because the objects are closer in mass to each other, unlike our solar system, where the sun outweighs the largest planet by a ratio of 1,000 to 1. A case in point is our nearest stellar neighbour, Alpha Centauri, a triple system 4.3 light-years away led by a star much like our sun. The three stars in the system are known simply as Alpha Centauri A, B and C. Alpha Centauri B orbits A at a distance approximately equal to that between the sun and Uranus. The third star—also known as Proxima Centauri because it is one-tenth of a light-year nearer to our solar system and is technically the closest of all stars to us—has a gigantic million-year orbit around A and B. In a gravitational sense, Proxima feels the combined masses of A and B as a single point. (A somewhat similar situation exists in our own solar system: Earth and moon share a common centre of gravity that is closer to the Earth's surface than its centre, and that gravitational focus, rather than the Earth's centre, is the point orbiting around the sun.)

Star A is slightly more massive than our sun and about 30 percent brighter. Its nearest companion, Alpha Centauri B, has about 90 percent of our sun's mass but only 40 percent of its luminosity. If a planet were orbiting Alpha Centauri A at the same distance that Earth orbits the sun, conditions would, in many ways, be like those we experience. Star A would appear the same size as our sun but would feel distinctly warmer. Life might be clustered closer to the planet's poles. More likely, an Earth-sized world at this distance from such a star would have surface conditions like Venus.

Alpha Centauri A and B revolve in elliptical 80-year orbits around a common centre of gravity. When the stars are closest, they are roughly a billion miles apart, a little more than 10 times the Earth-sun distance. At their farthest, they are 3.3 billion miles apart. From our hypothetical planet in orbit around A, sun B would look like a dazzling point of light most of the time. At its nearest, it would be a small, brilliant disc. Whenever B was in the night sky, the landscape of A's planet would be illumi-

How is it that the sky feeds the stars?
LUCRETIUS

The tattered remains of a star that exploded about 9,000 years ago are seen today splashed across the backdrop of the southern constellation Vela. This is the nearest and most spectacular example of a supernova remnant. The explosion must have illuminated the Earth's night sky nearly as brightly as a full moon. Yet all that remains of the star are these filaments and a dense cinder, called a neutron star, no wider than a small city. As the shock waves and gas from a supernova explosion push outward into space, they may meet a nebula —a cloud of gas and dust—and trigger its collapse, creating new stars and renewing the cycle of birth and death in the cosmos.

67

*Not only would a planet of a double sun, **right**, have two sunsets and sunrises each day, but its climate would oscillate to far greater extremes than the Earth's. Swinging around two suns increases the odds that a planet will be a wasteland like the one depicted here. Further inhibiting the emergence of life forms are changes in luminosity of stars, in either single or multiple systems. Despite its occasional flare-ups, **above**, our sun has proved to be a steady provider.*

nated to twilight level and only the brightest of the more distant stars would be visible to the unaided eye. One of those—our sun—would appear as a star in the constellation Cassiopeia.

The third star in the system, Proxima, is extremely faint and is 12,000 times the Earth-sun distance from the A and B pair. From a planet orbiting A, Proxima would be barely visible to the unaided eye, a dim starlike object that, if it were orbiting our sun at the same distance, would probably not have been noticed until after the invention of the telescope.

Could a planet like Earth exist in the Alpha Centauri system? For many years, this was considered highly unlikely, since B's varying distance from A would set up oscillating gravitational torques on the orbit of an Earthlike planet that would, over time, drastically modify its path around A. But these assessments have been reexamined, and it is now thought that a reasonably stable orbit may be possible for millions of years. But whether it would stay stable for billions of years, as in our solar system, is uncertain. The question is important when one considers the abundance of multiple-star systems. Some astronomers doubt that most multiple-star systems are born with planets in the first place be-

cause of the different speeds and motions of the nebular gas and dust during the birth of two or more stars close together. Pairs of stars separated by more than 10 astronomical units may be the only ones capable of having planetary families.

Multiple stars are systems within systems, but they are always arranged in increments of ones and twos: a single star orbiting around a pair, a pair of stars orbiting a single star, two pairs of stars orbiting about a common centre of gravity or, the usual situation, simply a pair of stars. The two stars in a binary system can be almost in contact with each other or up to a light-year apart. Of course, there are billions of single stars like the sun. All of them could have planets.

Planets of Other Suns

Astronomers and philosophers have debated the question of life on other worlds for centuries. The pendulum of opinion on the subject swung to its greatest extreme in the 18th century, when many astronomers were convinced that every star must have its attendant planets and that all those planets were inhabited. William Herschel, who discov-

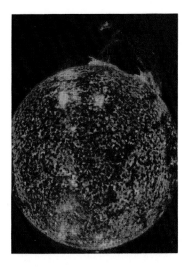

In comparison with our sun, **above**, *the red dwarf star Proxima Centauri,* **left**, *is a weakling that sheds only a feeble glow onto a hypothetical planet and companion moon. An intensive search for planets orbiting nearby red dwarfs is now under way, but because they are known emitters of flares proportionally larger than the eruption in the solar photograph, red dwarfs are considered unlikely to have habitable planets.*

ered the planet Uranus in 1781, even speculated that the sun was populated. He thought that the dark sunspots were openings in the sun's fiery atmosphere through which we might peer into the habitable realm below.

In the early 20th century, opinion reverted to the view that we may be alone in the cosmos. The sun's family of planets was regarded as a fluke of nature, wrenched out of the sun's gaseous body by the gravitational influence of a star that sideswiped the sun. Since stars are so far apart, such an encounter would be exceedingly rare. If our planetary system had been created in that way, it would probably be the only one in the Milky Way Galaxy.

By the middle of the 20th century, that hypothesis was discarded for a variety of theoretical reasons, the main one being that material torn from the sun would be far more likely to disperse or to fall back into the sun than to coalesce into planetary bodies. Today, the dominant theory suggests that during the birth of stars from clouds of nebulous dust and gas, planets form from abundant leftover material spread in a disc around the new sun, making planet birth a natural by-product of star birth.

Although there is convincing evidence that our solar system emerged in this manner, the concept would get a big boost if a planetary system could be detected orbiting another star. But the most powerful telescopes on Earth have been unable to spot a planet even as large as Jupiter. The difficulty is not so much the distance to other stars but, rather, the overwhelming brightness of a star compared with its planets. It is like trying to distinguish a penlight beside a floodlight.

Taking a more indirect approach, astronomers have searched for tiny wobbles in the paths of some nearby stars as they move almost imperceptibly through space against the background of more distant "fixed" stars. Such wobbles would be caused by the gravitational influence of large planets. If we could observe our sun's path from Alpha Centauri, the wobble induced by Jupiter would be as narrow as a line of type in this book seen at a distance of 100 miles. That kind of precision resolution still eludes astronomers, but a smaller star with a planet more massive than Jupiter would generate a detectable wobble. Tracking down these perturbations is extremely time-consuming because the star must be regularly observed for at least one complete orbit of its planet. Jupiter, for example, orbits

The largest planet possible is a gas giant about three times Jupiter's mass, top. More massive planets are compacted by their own weight to Jupiter's size or slightly smaller. Below Jupiter are Neptune, Earth, a red dwarf and the sun. Flares on a red dwarf can match the biggest bursts seen on the sun, which may mean that planets close enough to be warmed by a dwarf also get overheated now and then.

the sun once in 12 years. But the persistence of the handful of astronomers engaged in this work has paid off. After decades of monitoring, they have recorded wobbles in the paths of several stars, indicating that they are being orbited by superplanets many times more massive than Jupiter. However, this method of planet hunting pushes present equipment to the limit, leaving some uncertainty as to the exact nature of the companions. (Stars suspected of having companions are listed in the table of nearest stars at the back of the book.)

The suggestion that some stars have bodies more massive than Jupiter orbiting them leads to the question: where is the dividing line between a planet and a star? Of course, a planet is an object that is not massive enough to produce its own energy through thermonuclear hydrogen-fusion reactions—the most fundamental type of reactions that generate the energy stars radiate. But there is a complicating factor that, until recently, prevented a clear distinction between stars and planets.

Stars and planets are part of a hierarchy of the cosmos that includes a vast range of objects differentiated by their masses. Our solar system provides an inventory that demonstrates a portion of that gradation. At one end of the scale is the sun, a star over a million times the Earth's volume and 333,000 times its mass. At the other end are cosmic dust particles, visible only under a microscope, that constantly rain into the Earth's atmosphere at the rate of hundreds of tons a day (the silt in a house's eaves probably contains a minute amount of interplanetary material). Between these two extremes are worlds ranging from Pluto, less than 1 percent of the Earth's mass, to Jupiter, 318 times as massive as our planet. Overlapping the lower end of the regime are the planetary satellites, with Jupiter's Ganymede leading the pack that extends down to Mars's Deimos. Below this size are asteroids and comets. The major gap in the otherwise continuous size gradation in our solar system is the 1,000-fold increase in mass from Jupiter to the sun.

The sun is often referred to as a typical star, but actually, the average star in the galaxy is considerably smaller and dimmer. Astronomers call these plebeian stars red dwarfs because of their tint and small size. Red dwarfs constitute more than half the population of the entire Milky Way Galaxy. There are hundreds of billions of them, yet not a single one is visible without a telescope. Proxima Centauri is a red dwarf. It is about one-tenth the diameter of our

sun and one-tenth its mass, but it shines with only one twenty-thousandth the brightness. If it were to replace the sun, daytime on Earth would never be more than a sombre twilight.

The faintest red dwarf star known is LHS 2924, one-millionth the brightness of the sun and 28 light-years from Earth. It is detectable in large observatory telescopes. Just 8 percent of the sun's mass (80 times Jupiter's mass), LHS 2924 lies near the border line between true stars like the sun, which produce energy by fusing hydrogen into helium, and so-called brown dwarfs, which cannot ignite the thermonuclear furnace and merely generate a dull glow through compression—the mechanism responsible for Jupiter's significant internal heat.

Brown dwarfs below the transition line—roughly 70 times Jupiter's mass—masquerade as red dwarfs for about 100 million years because of the energy provided by their slow gravitational contraction. But once they compress to 90 percent of the diameter of Jupiter, they stop shrinking and cool off, slowly dimming over the next few billion years. Despite predictions that there might be at least as many brown dwarfs in the galaxy as there are stars, not one has been detected with certainty. Even if the nearest brown dwarf is closer than Proxima Centauri, it would be far too dim to be seen directly in visible light. But because its weak radiance is almost entirely infrared, it could be detected by the most sensitive infrared telescopes. As of 1992, infrared searches have uncovered possible brown dwarf companions of low-mass stars, but they remain unconfirmed.

Suppose a brown dwarf of 50 Jovian masses replaced Jupiter in our solar system. The changes would be surprisingly minor. The greater mass might disturb Saturn's orbit, but all the other planets would ply their paths around the sun essentially unaffected. Since they are nearly the same size, the brown dwarf would appear as Jupiter does in the Earth's sky—merely as a bright star—but a little more brilliant than Jupiter because its own light would be added to reflected sunlight, which would give it an orange tint. It would probably have dramatically active cloud circulation, possibly like Jupiter's but more rapidly modified by currents welling up from hotter regions below.

Just as Jupiter has the largest satellite system (by mass), so we would expect the more massive brown dwarf to have an even bigger retinue. There is no reason a moon of a brown dwarf could not be as

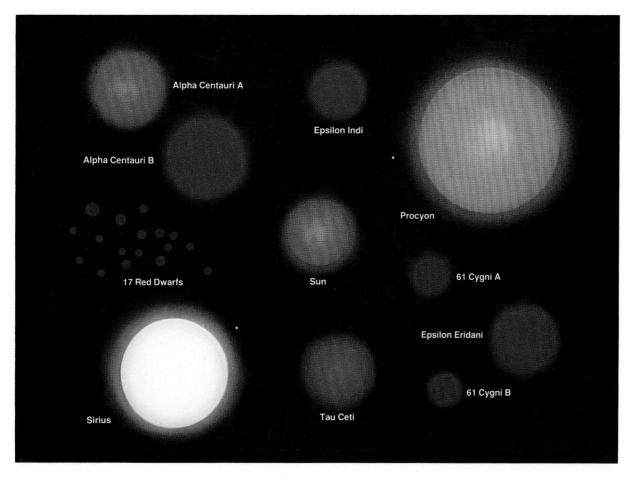

The comparative sizes and colours of all stars within 12 light-years of the sun are depicted here. Red dwarfs are the distinct majority, and only two stars, Procyon and Sirius, are decidedly larger than the sun. The small stars beside Procyon and Sirius are white dwarf companion stars.

large as Earth. Such a supermoon, in an orbit about the size of Europa's, would be heated to nearly room temperature on the inward-facing side. (All major planetary moons become gravitationally locked from youth so that one side perpetually faces the parent body.) That might provide an opportunity for life to emerge, though the specific environment is difficult to predict, since we have no experience with a world permanently warm on one side and frigid on the other. Nevertheless, there are attractive aspects to the idea. Brown dwarfs are probably as common as red dwarfs; they are simply too dim to be identified. Even if Earth-sized objects orbiting brown dwarfs are rare, lots of them ought to be out there, and they should be at least as interesting to visit as Europa or Titan.

Sun Death

The Milky Way Galaxy is the birthplace and graveyard of trillions of stars. No star lasts forever, but some have shone for 10 billion years and will continue to do so for 100 billion more. Ubiquitous red dwarfs have the greatest life spans because they cook their hydrogen fuel slowly at low temperatures. A star like Proxima Centauri will radiate at its current rate for at least 200 billion years—more than 10 times longer than the present age of the universe. At the opposite extreme are blue supergiants such as Rigel and Deneb. These massive stars live fast and die young. They flood the galaxy with 50,000 times the sun's radiance for a few million years, then blast themselves into oblivion. A Deneb or a Rigel is a conspicuous beacon even when seen from 1,000 light-years away. A star the brightness of the sun becomes lost in the stellar throng at distances greater than 50 light-years.

But regardless of their mass and projected life spans, all stars are fired by the same mechanism that makes the sun shine. In essence, the sun is a giant nuclear furnace. At its core, 655 million tons of hydrogen are fused into 650 million tons of helium every second, at a temperature of 27 million degrees F. The missing five million tons of matter are

71

DISTANCES IN THE COSMOS

More than a dozen different methods are used to measure distances in the universe. The main ones are given in the table. Among the various techniques, Cepheid variable stars are the key distance calibrators. The intrinsic brightness of each Cepheid is directly related to its rhythmic oscillations in luminosity. For instance, a Cepheid with a long period of oscillation has high intrinsic brightness, like a 100-watt light bulb, whereas a short-period Cepheid would be a 40-watt bulb, and so on. Comparing the real luminosity, or wattage, with the apparent brightness reveals the distance. Fortunately, Cepheids are typically more than 1,000 times brighter than the sun and can be seen across vast distances.

The upper-left image on the **facing page** is a CCD (still video) picture of the galaxy IC 4182, taken with the Palomar 200-inch telescope. (The bright spear is an overexposed star.) The boxed area is seen in more detail in the upper-right CCD image from the Hubble Space Telescope. The inset box in that picture is shown enlarged in the two lower images, taken with the Hubble five days apart. The marked Cepheid has gone from maximum to minimum in that period.

These observations, made in 1992, are important because IC 4182 is the nearest galaxy in which a type Ia supernova has occurred (in 1937). The discovery crucially links, for the first time, these two methods. The results suggest that IC 4182 is 16 million light-years away and offer better calibration of other galaxy distances.

Distance	Objects	Method	Technique or Assumptions	Accuracy
up to 50 AU	planets, asteroids and comets	radar	powerful radar aimed from Earth is bounced off object; round-trip time gives distance	can be accurate to a few feet
4 to 200 light-years	nearby stars	heliocentric parallax	photos of star taken when Earth is on opposite sides of its orbit; shift of star compared with background stars yields distance	accurate to 1 light-year for stars within 30 light-years; loses precision rapidly after 100 light-years
100 to 200,000 light-years	stars in our galaxy and Magellanic Clouds	main-sequence fitting	a star's spectrum often tells where it fits on main sequence and thus its true luminosity (see diagram page 77)	considered accurate within plus or minus 10%, though many giant stars are tricky to assess
to 20 million light-years	nearby galaxies	Cepheid variable stars	true luminosity of Cepheids is directly related to the star's period of variability (see text)	regarded as the most important "standard candle" in the universe
to 100 million light-years	nearby galaxies	blue supergiant stars	the very brightest blue supergiants in a galaxy are about the same luminosity as those in our galaxy	not as accurate as Cepheids, but blue supergiants are brighter
to 200 million light-years	nearby galaxies	globular clusters	the brightest globular clusters in a galaxy are about the same for all galaxies	not as accurate as above methods, but useful as a cross-check
to 300 million light-years	nearby galaxies	type Ia supernovas	type Ia supernovas are all believed to have nearly the same maximum luminosity	because Ia supernovas are brighter than anything else in a galaxy, this method reaches farther out
to 300 million light-years	galaxies	galaxy brightness	several techniques are used to estimate a galaxy's true luminosity	distance is essentially an educated guess
to 3 billion light-years	galaxy clusters	brightest galaxies	the largest and most massive galaxies may be similar from cluster to cluster	useful when galaxies are so faint and remote that no other method above works
100 million to 10 billion or more light-years	remote galaxies	redshift	because of the universe's overall expansion, the faster a galaxy is receding, the farther away it is (pages 107-12)	despite decades of work, different research teams often disagree by 50% or more on final estimated distances

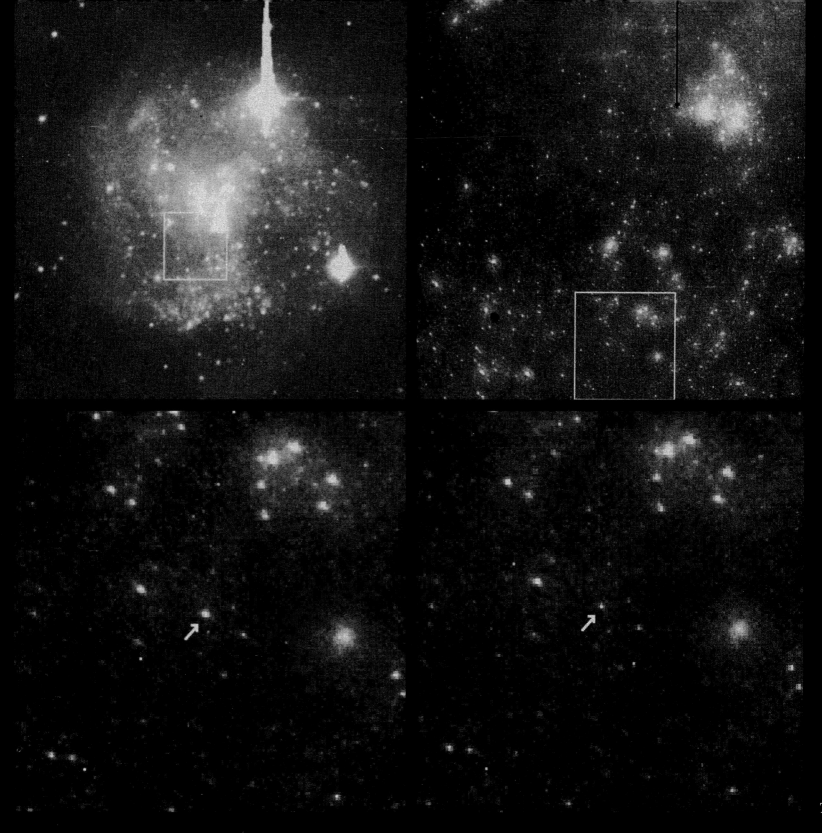

The shattered outer layers of a
star once similar to the sun
have been blown into space as
an expanding bubble of gas,
creating a planetary nebula
known as the Helix. The
remnant star at the focus of
the nebula provides the energy
for the cloud's brilliantly
fluorescing gas. In a few
million years, the nebula will
disperse, seeding the galaxy
with carbon-enriched gas for
new generations of stars.
Meanwhile, the central star will
decline into a white dwarf.

converted into 400 trillion trillion watts of energy in the process. After a tortuous trek lasting up to a million years, the core-generated energy works its way to the sun's surface where it is radiated into space, mostly as light. The sun's steady energy output is crucial for life on Earth. An abrupt change of only a few percent in the sun's production would vastly alter global climatic conditions. Apparently, the sun has never had a serious power outage. However, it will not always be so. In five or six billion years, a major disruption is inevitable.

The midlife crisis will be triggered by a depletion of hydrogen at the core. Starved for fuel to stoke its nuclear furnace, the sun will face an energy crunch. The thermonuclear reactions will then be transferred to a shell around the core, where hydrogen will still exist. The core will contract, which will heat up the surrounding layer of burning hydrogen, accelerating the reactions and producing more energy. Rather than dimming, the sun will become brighter.

On Earth, an increase in temperature will register, perhaps only a fraction of a degree a year, but in time, the enhanced solar radiation will melt the polar icecaps and make the equatorial regions intolerable. The transition will be slow. If, for example, the process had begun at the time the Egyptians were constructing the pyramids, the changes would be detectable today by instruments but would otherwise go unnoticed. However, global climate alterations of disastrous proportions would loom several millennia in the future.

Over hundreds of centuries, escalating quantities of energy being pumped out of its core would puff up the sun like an inflating balloon. Its diameter would double, triple, then quadruple. The energy output would accelerate as the core became hot enough to burn helium, which fuses to carbon, forcing the outer regions even farther into surrounding space. In a few million years, the edge of the sun would reach the innermost planet. Mercury has always been a dead cinder of a world, but at this point, its rocks would vaporize in the heat. Temperatures above the boiling point of water would extinguish all life on Earth. The oceans would vaporize to form a stifling atmospheric blanket that would be augmented by the noxious output from volcanic activity induced by rising surface temperatures.

The sun's expansion would continue for millions of years until the Earth's sky would be almost filled by the deep red distended globe with a glowing heart—the engine of the doomsday scenario—only partly concealed by the bloated outer layers. At that stage, the sun would be a full-fledged red giant star, thousands of times its original brightness. Earth might survive the holocaust, but only as a frazzled chunk of slag. What would happen next is less certain, but one widely accepted theory suggests that the immense energy flow from the red giant would act as a stellar wind, ejecting a steady stream of dust and gas. The gas would then form an expanding envelope around the star, eventually amounting to 10 percent or more of the star's mass.

Finally, 100 million years after the crisis began, the energy-producing core of the red giant would exhaust its nuclear fuels and collapse into a dense lump—a white dwarf—a stellar corpse radiating white-hot light due to compression. The white dwarf would have roughly three-quarters of the sun's original mass compressed into a body the size of Earth, its atoms crushed by gravity to a state where atomic nuclei swim in a dense sea of electrons. A teaspoonful of white-dwarf material would weigh about five tons. Other stars have met the same fate; hundreds of white dwarfs have been discovered within a few dozen light-years of the sun.

During its formation, the white dwarf produces a blast of stellar wind even more energetic than that which emerged from the red giant. This has a snowplough effect, pushing the previously expelled material into a discrete bubble about one light-year across. Like the filament in a light bulb, the dwarf's radiation lights up the surrounding gas, creating a distinct sphere, or doughnut, called a planetary nebula that can be seen for thousands of light-years —the star's last gasp. (Because these objects were described by 18th-century astronomers as being similar in appearance to Uranus and Neptune, they were called planetary nebulas.) After about 40,000 years, a planetary nebula disperses.

The white dwarf that will represent the sun's senile old age will drift among the stars in the same path it followed around the galaxy in its more robust youth. To the cinder that was once Earth, if it still exists, the new white dwarf sun will be a dazzling celestial diamond shedding a twilight-level glow, but virtually no heat, on a frozen wasteland. Over 30 billion years or so, the sun will slowly cool, like a dying ember, until it no longer radiates energy.

The final corpse of the sun will be a black dwarf about the Earth's size and 200,000 times its mass. So far, there is no evidence of any black dwarfs. The

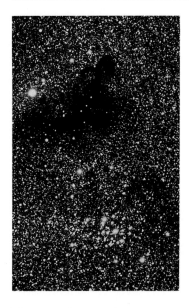

Masquerading as a hole in space, a nearby cloud of cosmic gas and dust several dozen light-years across completely blocks light from more remote stars in our galaxy. Sooner or later, such clouds collapse to create a stellar birthplace—a celestial womb where a cluster of stars, like the one nearby in this scene, will be born. Astronomers estimate that one new star per year, on average, arises somewhere in the galaxy in this fashion. But more stars die than are born. In tens of billions of years, the galaxy will be a much dimmer place than it is today.

THE COLOUR, SIZE AND BRIGHTNESS OF STARS

Stars are not created equal. They inherit different amounts of material from the nebulas in which they are born. Super-heavyweight stars are 100 times the sun's mass. Minimum stellar mass is 8 percent of the sun's. A star's mass dictates how bright it will be, how long it will live, its temperature and its size. The diagram, right, provides this information for several dozen stars that represent a cross section of the stellar population. Although it requires some explanation, the diagram is a powerful tool and can be used to plot stars other than those indicated.

Astronomers have determined the intrinsic brightness and temperature of thousands of stars, several hundred of which are shown as black dots. The background colour represents the actual colours of the stars. The coolest stars are red, the hottest blue. The range from the extreme right of the diagram to the extreme left is 2,000 degrees F to 100,000 degrees F. Astronomers prefer to use the long-established spectral type for temperature classification. There is no significance to the letters in the letter-and-number spectral classes; it is a modern version of an old system. Each letter represents an arbitrary class. Within a spectral class, there are numeric subdivisions. In the B class, for example, a B0 star is a little hotter than a B1, which is hotter than a B2, and so on, to B9. Then comes A0, A1, A2, et cetera. Spectral class is determined through exami-

nation of the spectrum of star-light. Luminosity is calculated by measuring a star's apparent brightness and ascertaining its actual brightness based on its distance from Earth. If the distance is not known, the spectrum is compared with that of similar stars whose properties are known.

Once a star's spectral type and luminosity have been established, it can be plotted on the diagram. When hundreds of stars are noted in this fashion, a definite clumping occurs in a gentle S-shaped curve from upper left to lower right. Astronomers call this the main sequence. Ninety percent of the stars in the galaxy are found here. The main-sequence stage is the stable hydrogen-burning period of a star's life. An individual star does not evolve up or down the main sequence but stays essentially in one place until its hydrogen fuel is exhausted and an abrupt change in internal energy production occurs. At this point, it ceases to be a main-sequence star and will move to the upper right of the diagram. In the case of our sun, it will eventually climb to a position about midway between Aldebaran and Antares. Following the red-giant phase, a quick decline in luminosity will send it to the middle left, then down to the white-dwarf zone.

After a star contracts from its mother nebula and begins radiating energy through thermonuclear reactions, its brightness and temperature will correspond to a fixed location in the main sequence, depending on its mass. Stars the same mass as our sun are yellow and will last about 10 billion years in the main sequence. Stars one-quarter the mass of the sun are red and have only 1 percent of

the sun's luminosity, but they will remain on the main sequence more than 100 billion years. Conversely, stars of greater mass are hotter and brighter and have shorter main-sequence lives.

Thus a glance at the diagram yields the stars' temperatures, sizes, masses and main-sequence lifetimes. Stars off the main sequence are another story. Their luminosities, temperatures and sizes are indicated, but their masses and the length of time they will exist in their current state vary from one individual to another. For example, the red giant Mira is roughly the same mass as the sun, whereas the red supergiant Betelgeuse is 18 times the sun's mass and Rigel is 30 times greater. A further complicating factor is that the stars contain different ratios of elements (depending on the initial mix in the genesis nebula), although the primary constituents are always about three-quarters hydrogen and one-quarter helium, with a few percent of the mass taken up by all the other elements. These slight variations account for the spread of stars along the width of the main sequence.

Spectral Type	Surface Temperature (degrees F)
O5	72,000
B0	50,000
B5	27,000
A0	17,000
A5	14,800
F0	12,800
F5	11,400
G0	10,400
G5	9,400
K0	8,300
K5	7,000
M0	5,800
M5	4,500
M8	3,800

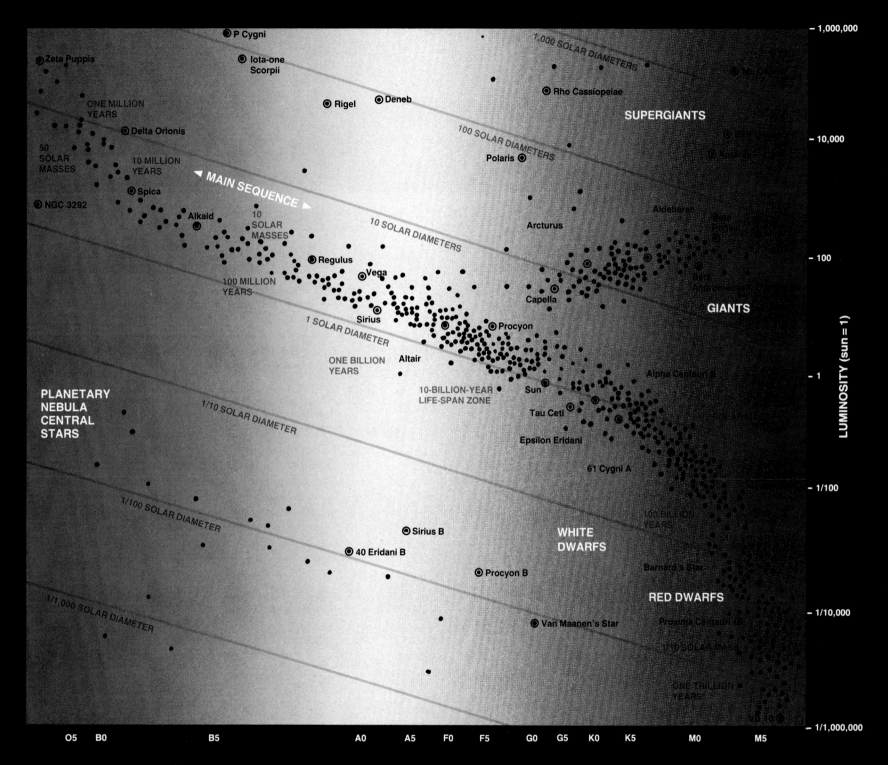

PLANETARY
NEBULA
CENTRAL
STARS

SUPERGIANTS

GIANTS

WHITE
DWARFS

RED DWARFS

MAIN SEQUENCE

50
SOLAR
MASSES

ONE MILLION
YEARS

10 MILLION
YEARS

100 MILLION
YEARS

ONE BILLION
YEARS

10-BILLION-YEAR
LIFE-SPAN ZONE

100 BILLION
YEARS

ONE TRILLION
YEARS

10
SOLAR
MASSES

1,000 SOLAR DIAMETERS

100 SOLAR DIAMETERS

10 SOLAR DIAMETERS

1 SOLAR DIAMETER

1/10 SOLAR DIAMETER

1/100 SOLAR DIAMETER

1/1,000 SOLAR DIAMETER

1/10 SOLAR MASS

P Cygni

Iota-one
Scorpii

Zeta Puppis

Deneb

Rigel

Rho Cassiopeiae

Delta Orionis

Polaris

Spica

NGC 3292

Alkaid

Aldebaran

Regulus

Vega

Arcturus

Sirius

Capella

Procyon

Altair

Sun

Alpha Centauri B

Tau Ceti

Epsilon Eridani

61 Cygni A

Sirius B

40 Eridani B

Procyon B

Barnard's Star

Van Maanen's Star

Proxima Centauri

VB 10

LUMINOSITY (sun = 1)

— 1,000,000

— 10,000

— 100

— 1

— 1/100

— 1/10,000

— 1/1,000,000

O5   B0   B5   A0   A5   F0   F5   G0   G5   K0   K5   M0   M5

SPECTRAL CLASS

77

Throughout its life, a star wages a battle between the force of gravity squeezing in and the outward flow of energy from its core. Massive stars, with greater core pressures, convert hydrogen to helium at a far greater rate than the sun and are therefore proportionately larger and brighter. In a few million years, the core of a massive star becomes clogged with helium. Gravitational compression escalates the internal temperature to burn the less efficient helium, rather than hydrogen.

The star gets bigger and hotter, and the stellar furnace is cranked up to higher and higher temperatures. The star bloats to a vast red giant, possibly wider than Jupiter's orbit. Heavier elements are created and promptly burned: carbon, nitrogen, oxygen, magnesium and, finally, iron. But iron cannot be used as stellar fuel. It is as inefficient as stones in a fireplace. The star becomes rotten at the core, choking on the waste products of its fuel consumption. Now, a lethal death spiral seals the fate of the great star. Energy production ceases, and gravity takes over. The star's core implodes, while the outer layers reach temperatures of billions of degrees and erupt in the explosive fury of a supernova. Radiation floods into surrounding space and is visible millions of light-years away.

In some cases, a giant star going supernova releases as much energy as the radiation of 10 billion ordinary stars. From a distance, the giant star's death appears as a new star bursting forth where nothing was seen before. The supernova dwindles a few years after the eruption, but its effects are felt for centuries. The shock from the explosion continues to surge into the galaxy, a tidal wave of energy in an ocean of space. But interstellar space is not entirely empty. An exceedingly tenuous mixture of gas and dust gradually slows the expansion. If the supernova erupts within a few dozen light-years of a region where the gas and dust are concentrated in a discrete cloud, the shock wave wraps around the nebula, compressing it into a more compact object—like ghostly cosmic hands packing a celestial snowball. It is from such compressed nebulas that stars and their families of planets are born.

Supernova shock waves are thought to be a primary initiator of stellar formation. Moreover, the ashes of the dead star—those elements cooked up before and during the eruption—enrich the nebulas, making the element ratios in the new generation of stars different from the old. All the metals on Earth were forged in the fires of massive stars.

universe is not old enough for any of the white dwarfs to have completed the slow withering death that appears to be the fate of our sun.

Supernova: A Giant Star's Doom

Stars more massive than the sun end their lives in a spectacular fashion. It happens so abruptly, it is as if the galaxy is littered with cosmic time bombs. About once every half century, a giant star annihilates itself in a supernova explosion. Most of the eruptions go unnoticed from Earth, concealed by the dust and gas between the stars in the arms of the Milky Way Galaxy. But every few centuries, a star several thousand light-years away destroys itself in a supernova. The blast appears as a dazzling new star, bright enough to be visible in full daylight.

The detonation produces a fireball a million times hotter than the surface of the sun. Not only are such cosmic infernos spectacular, but the processes associated with them are believed to be a critically important factor in the birth of stars and galaxies and in the creation of the elements necessary for life. All stars do not become supernovas. In fact, most do not. Only those above six solar masses have the potential to go out in a blaze of glory.

Elements heavier than iron—lead, gold, uranium—were created in the fire storms of supernovas. Without supernovas, elements heavier than carbon would be exceedingly rare and planets like Earth could not exist. In a universe without supernovas, it might be impossible for life to develop.

Although millions of supernovas must have illuminated our galaxy since the first stars were born, they are not common. The brightest supernova since the invention of the telescope erupted in the Earth's sky on February 23, 1987. At maximum luminosity two months later, it rivalled some medium-bright stars—easily seen, but inconspicuous to anyone not looking for it. The explosion had occurred 170,000 years earlier in our nearest neighbour galaxy, the Large Magellanic Cloud, 170,000 light-years distant. A supernova seen in our own galaxy in 1604, about 25,000 light-years away, was as luminous as Jupiter. One observed in 1572 was only 10,000 light-years distant and shone as brightly as Venus.

There are records, mostly in Chinese chronicles, of supernovas occurring before that. One appeared in 1054 A.D. and was called a guest star by Chinese skywatchers. Today, in the exact place the Chinese records indicate, we see the Crab Nebula, a three-light-year-wide cloud of contorted star material blasted into space by a supernova.

A little more than nine centuries later, in 1968, astronomers learned that the Crab Nebula is a remarkable cosmic crucible. Deep within the nebula's twisted filaments of star-stuff lies a pulsar, an object so bizarre that even though it was predicted in the 1930s, most scientists ignored the theoretical basis for its existence. The pulsar is a whirling ministar flashing an intense, magnetically focused beam of energy like an emergency vehicle's beacon. No wider than a small city yet 500,000 times the Earth's mass, a pulsar is the superdense imploded core of the star that erupted as a supernova.

Pulsars are commonly called neutron stars because of the nature of the atomic particles that make up these ultracompact objects. A neutron star is so alien to human concepts that it almost defies description. It is difficult to imagine how a mass of 500,000 Earths can be crammed into a sphere 10 miles wide. Suppose a balance scale could be built to hold a teaspoonful of neutron-star material. What would equalize the balance? Obviously, this is heavy stuff, so let's try a railroad locomotive. No, too light. Next, a half-million-ton oil tanker. No, that

The Crab Nebula, a contorted cloud of gas and stellar debris 5,000 light-years from Earth, is a celestial tombstone marking the death of a massive star seen from Earth as a supernova in the year 1054. Deep within the three-light-year-wide cloud is a neutron star, the original star's final remnant. The spinning neutron star—a pulsar—was detected from Earth in 1969. **Facing page:** In about five billion years, the sun, as a full-fledged red giant star, will bloat to 100 times its present diameter, reducing Earth to an airless chunk of slag that may be consumed entirely before the giant sun's outer layers are blown off in a planetary nebula similar to the Helix (page 74).

would not even budge it. No man-made object is massive enough to tip the scales. It would take a two-mile-high mountain to equal the mass of this tiny amount of unearthly material.

It is easy to understand why, prior to the actual detection of a neutron star, some scientists preferred to dismiss the concept as a theoretical curiosity unrelated to the real universe. But neutron stars do exist, and their mighty magnetic fields focus energy beams that flash as pulsars when they are young and rapidly rotating. Like whirling tops, pulsars eventually spin down. Their pulsing, which gains much of its energy from twirling the magnetic field up to near light-speed, dies out.

A deactivated spun-down neutron star would be a less formidable body to visit than one that is rapidly spinning and soaked with radiation in its pulsar phase (the Crab Nebula pulsar spins 30 times per second). Even so, there are awesome powers associated with older neutron stars. A several-billion-year-old neutron star would spin completely around about once a minute, still a dizzying rate for something the size of a small city. At close range, it would resemble a white-hot ball bearing. Although in theory a spacecraft could approach such an object, a landing could never happen. The enor-

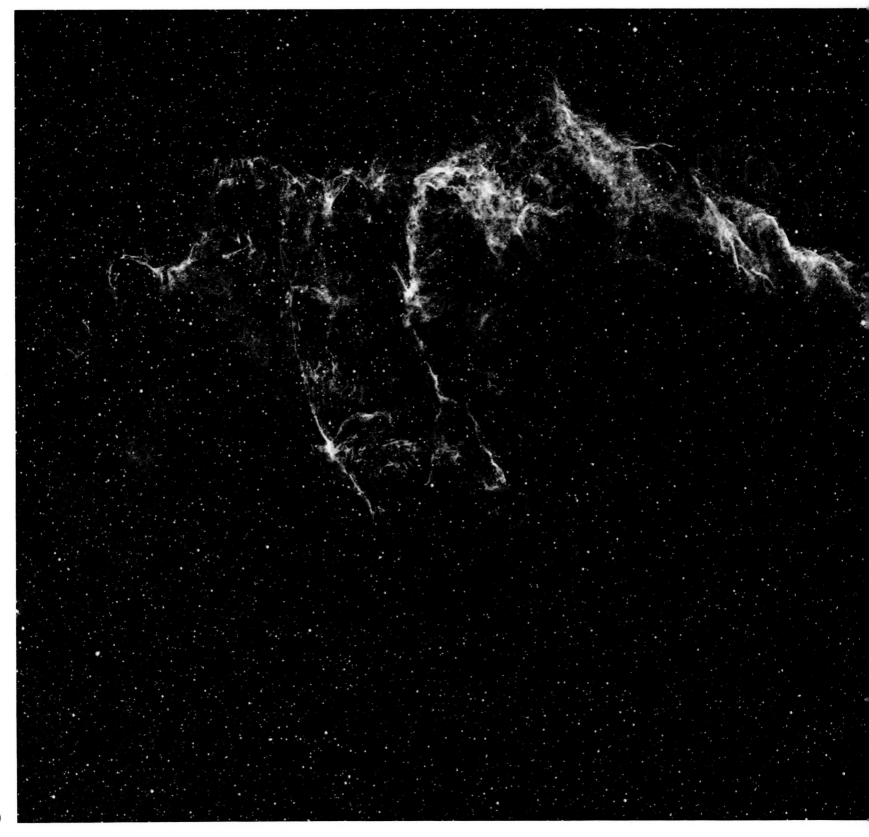

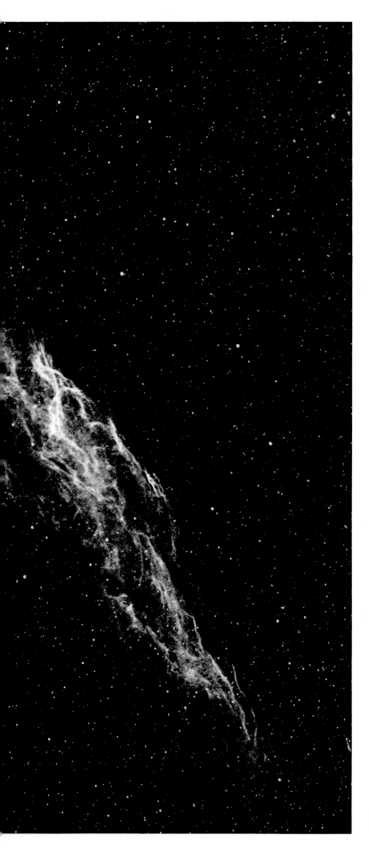

mous surface gravity would instantly crush a spaceship and its occupants, flattening them into a puddle of subatomic particles. Mountains on a neutron star would be measured in millimetres, not miles, even though they would contain the same amount of matter as mountains on Earth.

Following the discovery of the Crab Nebula pulsar, astronomers sought other pulsars associated with supernova remnants. The debris of past supernovas litters the sky like static puffs of smoke, but despite the detection of hundreds of pulsars, only two have been found embedded in matter expelled by a supernova—the Crab Nebula pulsar and one at the centre of a large tattered ring of gas in the constellation Vela. The supernova that created the Vela pulsar occurred between 6,000 and 11,000 years ago, an approximation based on estimates of the supernova remnant's velocity as it hurtles away from the site of the explosion. Astronomers calculate that the Vela pulsar is just over 1,000 light-years distant, four times closer than the Crab Nebula pulsar. At that distance, the Vela supernova must have been nearly as bright as the full moon, far more dazzling than Venus.

Since the Vela supernova lit up the sky between 4000 and 9000 B.C., humans certainly observed it. If it occurred closer to 4000 B.C., its appearance may have been recorded. Around that time, the Sumerians were the leading civilization—the first to develop cities, a system of government and a form of writing. From their settlements in Mesopotamia (modern-day Iraq), the supernova would have appeared close to the southern horizon. The new star would have shone 1,000 times brighter than Venus, easily casting shadows. For months, it would have been a diamondlike second sun by day and a mysterious stellar beacon by night.

There has not been a supernova visible in our galaxy since Kepler's observation nearly four centuries ago, although the February 1987 supernova that erupted in the Large Magellanic Cloud was just beyond the outskirts of the Milky Way. Still, it would be a treat to get a real dazzler in our lifetimes. The odds are not good, though: in the past millennium, only five have rivalled the bright planets.

Black Holes: Gravity Whirlpools

Gravity, the master architect of the universe, has always fascinated astronomers. Orbits, masses, densities, velocities—all are controlled by gravity.

The Veil Nebula, one of the heaven's most delicate structures, is a 20-light-year-long segment of a nearly complete ring of tattered material blown off during a supernova that occurred about 30,000 years ago. No remnant object, such as the pulsar at the centre of the Crab Nebula, has ever been found.

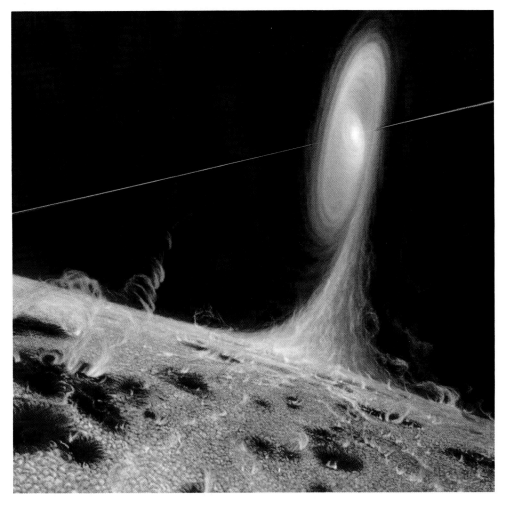

surmised, would be invisible to distant observers.

Little could be added to Michell's concept until the 20th century. In 1916, shortly after Albert Einstein formulated the theory of general relativity, his German colleague Karl Schwarzschild realized it predicted that a star of sufficient compactness and density would be crushed by its own gravitational pull. Only the gravitational field would remain; the object itself would disappear. For a celestial body of the sun's mass, he calculated the critical diameter at about four miles. Once a solar-mass body collapses past that size, it has dug itself a black hole in space and disappears into it. The critical size, called the event horizon, marks the black hole's boundary. It varies in direct proportion to the mass of the object that collapsed to create the hole.

A black hole with the Earth's mass would have an event horizon the size of a golf ball. But our planet cannot become a black hole. Neither can the sun. They are not massive enough for gravity to have the last word. Earth is as compressed as it will ever get, and the sun will not shrink past the white-dwarf stage. But stars more than 1.4 times the sun's mass can be compressed to incredible densities as neutron stars. Neutron stars have their limits too. Stars over four solar masses cascade past the neutron-star stage to the ultimate doom of a black hole. A massive star is just a black hole waiting to be born.

A black hole forms in the same type of supernova fireball that creates neutron stars. If the exploding star exceeds about 10 times the mass of the sun, the remnant imploding core may be more than four solar masses. If it is, gravity takes complete control, crushing atomic particles into each other with such fury that nothing can stop the infall. The matter that creates the black hole disappears, leaving only the gravitational field. Like Alice in Wonderland's Cheshire cat, all that remains is the disembodied grin of its gravity. From afar, the black hole's gravity has the same effect on objects in space as it did when its matter existed, but closer in, the gravitational force soars, becoming so great that it prevents the escape of light.

Although several massive black holes are suspected to exist at the cores of most galaxies, only four stellar-mass black holes have been identified within our galaxy. Cygnus X-1, the first, was discovered in 1971 when its intense x-ray radiations were recorded by the first orbiting x-ray observatory. (The "sighting" had to await orbiting instruments because the Earth's protective atmosphere screens

In 1784, John Michell, an English clergyman, amateur astronomer and geologist who became famous for his invention of the torsion balance and for the creation of seismology as a science, attempted the first prediction of the ultimate extension of gravity's power. Using Sir Isaac Newton's formulas, Michell calculated that an object must attain one twenty-five-thousandth the speed of light to escape the Earth's gravity. To break away from the sun's gravity, a particle would have to achieve one five-hundredth the speed of light (light's speed, 186,282 miles per second, was known with reasonable accuracy at that time). Michell extended his musings one step further in a letter to Cambridge physicist Henry Cavendish. He suggested that if the mass of the sun were to be increased by a factor of 500, the escape velocity would equal the speed of light. He concluded that "all light emitted from such a body would be made to return toward it by its own proper gravity." The more massive star, he

out all celestial x-rays.) The x-ray source coincides with the position of a star known as HDE 226868, a giant blue star about 27 times the sun's mass at a distance of approximately 11,000 light-years.

Spectroscopes on Earth-based telescopes revealed that the blue giant approaches then pulls away from Earth over a 5½-day cycle, indicating that it is in a 5½-day orbit around an invisible object. By analyzing this orbit, astronomers calculated that the object must be roughly 15 times the mass of our sun. Since the maximum for a neutron star is four solar masses, there seems to be no alternative to a black hole.

The two objects in the Cygnus X-1 system are one-fifth of an astronomical unit (the distance from Earth to the sun) apart, and the black hole is swallowing up matter from the blue star at the rate of 100 billion tons a day. As it plunges into the hole, the material heats up to over one billion degrees F, resulting in the x-ray emissions that first drew astronomers' attention. Matter swirling around the black hole forms a flat disc at the hole's equator (the momentum of the original star's rotation has been preserved, giving the hole an equator and poles). These accretion discs, as they are called, are thought to exist around all high-density objects that are pulling matter from their surroundings—black holes, neutron stars and white dwarfs. When such high-density objects are coupled in close binary systems, prodigious amounts of high-energy radiation are generated as the compact object heats, then gobbles up material from its neighbour.

The other strong black-hole candidates in or near our galaxy are: LMC X-3, a six-solar-mass blue main-sequence star orbiting a suspected nine-solar-mass black hole in 1.7 days; A0620-00 in Monoceros, a sun-sized star orbiting an eight-solar-mass dark companion in eight hours; and V404 Cygni, another sun-sized star orbiting a suspected 10-solar-mass black hole in 6.5 days.

Since their discovery, black holes have posed an intriguing question: what are they like close up? Because of the intense accretion-disc radiation, it would be lethal to venture anywhere near one like Cygnus X-1. However, a 10-solar-mass black hole without a companion star could be approached within a few million miles. Lacking the food supply from a neighbouring star and the energy emissions that go along with consumption, the hole would be truly black, impossible to see. But it would still have the same gravity as a star 10 times the sun's mass.

Explorers in the vicinity of this object would need sensitive instrumentation to pin down its precise location. A black hole of 10 solar masses is just 40 miles in diameter. At this size, even if there were something to see, it could be detected only at close range.

As it approaches the black hole, the exploration ship establishes an orbit around it at a safe distance, then drops instrumented probes with accurate clocks into inward-spiralling orbits. The probes report nothing unusual, reacting exactly as if they were in orbit about a star of similar mass. Only when the probes close in to less than the Earth-moon distance does the power of the hole become unmistakable. From here, the hole's gravitational force simultaneously compresses and stretches any normal substance as if it were modelling clay. At 100 miles, the clocks on the space probes appear to run slow by about 15 percent. This time warping is caused by the hole's enormous gravitational field pulling on the signal being sent back by the probe. The effect increases until at the surface of the hole, any object, whether spiralling down or going straight into the hole, is accelerated to the speed of light before finally plummeting in. Once the probes cross the event horizon, their returning signals, even at the speed of light, do not escape. The hole is a gravity whirlpool, a tunnel into the substratum of space from which there is no return.

Anything that enters a black hole leaves the universe—*our* universe at least. To our perceptions, it is gone. Nothing inside a black hole can communicate with the universe it left. No known force can break a hole's grip. A black hole is omnivorous, consuming anything—a one-way trap in time and space. Whatever crosses the event horizon is stretched spaghetti-thin, pulverized by gravitational tidal forces and sucked into the singularity, the black hole's heart.

Because the structure of black holes is not fully understood, there is no consensus about the ultimate fate of material consumed by these gravity whirlpools. Is it pumped out of the universe forever, or are there tears in the fabric of space-time where mirror objects—white holes—fountain the matter back into the universe at another time and place? Are black holes the key to the space-warp drives that prop up so much of science fiction? Many astrophysicists say none of these possibilities are likely. Black holes may simply be what they appear to be: the universe's ultimate abyss.

More powerful than 250 million suns, the supernova of 1987—the brightest seen in 383 years—was the astronomical event of the decade. Located in our nearest neighbour galaxy, the Large Magellanic Cloud, the exploding star came close to outshining its home galaxy. It is seen near maximum brightness at lower left in this photograph. It remained visible to the unaided eye for most of 1987 and is now slowly fading. Astronomers expect to find a pulsar, the twirling collapsed core of the original star, when the brilliance of the blast subsides. The pinkish contorted cloud is the Tarantula Nebula, also seen on page 92. **Facing page:** *SS433 is one of the most remarkable objects in the galaxy. It may be a black hole being fed more than it can consume. Material from the large star pours onto the hole's accretion disc, and the overflow squirts out in two jets at an incredible 50,000 miles per second.*

The nucleus of the Milky Way Galaxy, **left**, viewed from a distance of less than one light-year, reveals a waltz of doom orchestrated by the laws of gravity. Three powerful black holes pivot around a super-massive (four-million-solar-mass) black hole that defines the maw of the galaxy. Eventually, the most massive hole will devour the others, adding their mass to its own. Abundant gas in the galactic core forms ghostly tendrils as it swirls into the holes. High-energy radiation released by the holes as they consume the gas has been detected on Earth, 25,000 light-years away, but the details shown here are still speculative. **Above**, an even more massive black hole near the centre of another galaxy is depicted. The puzzling polar jets are thought to be excess matter expelled over the hole's rotational poles.

The splash, the dash and the crock: three completely different galaxies. Our nearest galactic neighbour, the Large Magellanic Cloud, **far left**, is littered with stellar nurseries like the Orion Nebula (page 132). The largest is the Tarantula, at upper left, 10 times wider than the Orion Nebula. Some tattered rings of material are supernova remnants. The Large Magellanic Cloud may be a barred spiral, but its ragged shape makes it difficult to classify. Similarly, the peculiar elliptical galaxy NGC 5128, **left**, has dust lanes characteristic of spirals, suggesting that it may be the product of a collision of both types. The nearly edge-on spiral NGC 253, **above**, is slightly smaller than the Milky Way.

93

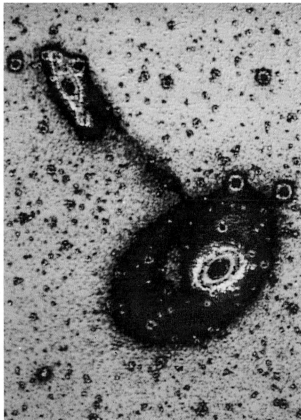

arm of the galaxy. Since the sun's velocity is twice that of the Orion arm density wave, we should be passing through it in about 10 million years.

A second theory to explain spiral arms suggests that the collision of large nebulas in the galaxy starts a chain reaction of star formation, supernovas and more star formation. The cycle continues for a few hundred million years during which the rotation of the galaxy twists the star-forming region into a long spiral shape. Both theories have their adherents, but there is still much to learn about how the immense star cities have become stabilized in their dazzling pinwheel structures.

Galactic Contrasts

Spiral galaxies are undoubtedly among nature's most elegant creations. Their graceful curving arms, traced by millions of giant blue stars and puffs of pink nebulosity, contrast with the bulging nucleus where 100 billion stars are nestled.

Compared with spirals, elliptical galaxies are rather dumpy. Ellipticals are star piles, vast spherical or football-shaped blobs of yellow and red stars.

Though not ugly, ellipticals certainly lack flair.

Why did two main types of galaxies form after the Big Bang? Why are they not all spirals?

Until the early 1980s, astronomers assumed that the manner in which the primordial gas clouds collapsed determined the type of galaxy that was born. But that idea is being seriously questioned. There is growing evidence that the genesis galaxies were of only one type: spirals.

In the new view, ellipticals are the wreckage of spirals that collided within a few billion years of their formation. Because the universe is growing larger due to continuous expansion since the Big Bang, it must have been much more compact and crowded 14 billion years ago when the galaxies were born—about 1 percent of its present size. But the same amount of matter was in the universe at that time. The newborn galaxies were almost rubbing shoulders, and collisions would have been far more common than they are today.

A galaxy collision would last for hundreds of millions of years. Once two spirals actually touch, they are likely doomed to loop ever closer to each other before finally merging, although because of the distance between them, individual stars would virtually never collide. Even a gentle merger would create an elliptical because collisions among the abundant gas clouds in the young galaxies would trigger star formation. Friction from the cloud collisions, plus dynamical friction from the gravitational interaction of the two great star cities, would rip apart the spiral arms and cascade the remains into a chaotic mass of stars—an elliptical galaxy.

Even a close miss would twist off much of a spiral galaxy's twirling arms, although the nuclear region would remain essentially intact. In such a gravitational tug-of-war between two continents of stars, the outrider stars would receive two sets of signals that would scramble their orbits. As a result, many billions of stars would be flung into intergalactic trajectories and lost from both galaxies forever.

Further evidence supporting the scenario is that spiral galaxies are almost never located among the denser galaxy clusters. Spirals are usually loners, those that avoided being torn apart when the universe was younger. Galaxy collisions are relatively rare today, although NGC 5128, a nearby galaxy, was recently determined to be the debris from two galaxies that collided two billion years ago. At least one of the originals was a spiral. Elliptical galaxies, it seems, are not born, they are made—heaps of stars from dismembered spirals.

Journey to the Galaxy's Centre

Out in the Orion arm of the Milky Way Galaxy, where the sun resides, there is one star for every 400 cubic light-years. The next spiral arm inward, 6,000 light-years from us, is the Sagittarius arm, which is more generously endowed than ours with stars and nebulas. From deep within the Sagittarius arm, the galactic landscape would appear noticeably brighter than the sky in our region of the galaxy. The Centaurus arm is next, about 8,000 light-years farther in toward the galaxy's nucleus. Emerging from the Centaurus arm, a galactic traveller would see a wall of stars ahead— the hub of the galaxy, the galactic central bulge. From this distance (less than 10,000 light-years), the core region has a rich golden glow, the combined light of billions of stars far more closely packed than anywhere in the spiral arms. Very few blue giants are seen. Instead, the brightest stars are red giants like Betelgeuse and Antares. The vast majority of stars are yellow and orange.

Unlike the most luminous parts of the spiral arm, the nucleus does not derive its brilliance from short-lived blue giants but, rather, by brute force from throngs of lesser suns whose combined light creates a dazzling scene, like hundreds of bursting skyrockets superimposed and frozen in time.

Two thousand light-years from the galaxy's centre, the nucleus is still dimmed by clouds of gas and dust that mingle with the stars. At 150 light-years from the core, a little more than twice the distance from our sun to the majority of the stars in the Big Dipper, the galaxy's nucleus can just be glimpsed. It appears as an intense concentration of light, almost as radiant as the sun, surrounded by countless stars, mostly yellow but highlighted by red giants. Thirty light-years away. The nuclear region now exceeds the brilliance of full sunlight, but still the heart of the galaxy is concealed from direct view by gas and dust illuminated by 60 million stars within a 30-light-year radius.

Veiled by gas and dust, the maw of the galaxy is still not visible even at a distance of three light-years. In a 400-cubic-light-year zone at the galaxy's nucleus—a volume in which only a single star would be found in the spiral arms—there are more than two million stars along with liberal amounts of churning gas and dust. There is no night here. The sky contains thousands of stars as bright as the planet Venus and dozens as luminous as the full moon. The average distance between stars is just 20 times the diameter of Pluto's orbit. At less than one-tenth of a light-year from the core, the heart of the Milky Way Galaxy reveals itself: a monstrous vortex of gas, dust and star-stuff swirling to its final resting place at the bottom of a black hole. Attending to the leftovers are smaller gravity whirlpools, all in a death-spiral dance around each other. It is a scene humans may never see. The entire region is an inferno of lethal radiation.

The guardian of the galaxy's central fortress is a four-million-solar-mass black hole. Such a massive hole dwarfs the 40-mile-wide hole in the Cygnus X-1 system. It is about 15 times the diameter of the sun, and its accretion disc extends out to a distance equivalent to the diameter of Jupiter's orbit. According to theory, material whirling into a giant black hole would not plunge in immediately. Rather, it would be partially repelled and heated by intense radiation from the zone around the hole. Matter here is so compressed and is accelerated to such enormous velocities by the hole's gravity that it

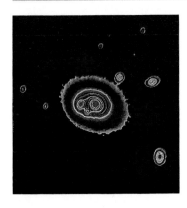

This false-colour electronic CCD image of the giant elliptical galaxy Abell 2199 shows galactic cannibalism in action. The overall bull's-eye pattern is the giant elliptical, its outer fringes indicated by the blue halo. (Black sky is shown brilliant green.) The two knots just to the left of the major galaxy's nucleus are a pair of galaxies beginning to be ingested. In about one billion years, they will lose their identities, their stars spreading throughout the large galaxy. More potential victims swarm on the outskirts of the swelling galaxy, flirting with almost certain demise.
Facing page: *Clash of the titans. Two colliding spiral galaxies in the constellation Corvus,* **left,** *begin to lose their individual identities during a billion-year encounter that will leave both severely distorted. The false-colour electronic image,* **right,** *reveals a pair of galaxies involved in a celestial sideswipe.*

THE ELECTRONIC UNIVERSE

The electronic and computer revolutions of the 1970s and 1980s have vastly altered the way astronomical data are obtained and processed. Fewer photographic exposures are being made each year as electronic detectors become ever more sophisticated.

The most widely used electronic detector is the charge-coupled device, or CCD. About the size of a postage stamp, **far right**, a CCD mounted at the telescope's focus acts like a miniature television camera. Light from celestial objects impinging on the CCD's surface is recorded on magnetic tape as electronic impulses. The CCD is 25 times more sensitive to light than photographic film is; it literally counts the photons of light the telescope delivers. Its only disadvantage is that it covers a smaller area of the sky than do photographic plates. What it does see, however, is recorded in exquisite detail. A CCD attached to a 24-inch telescope makes it perform like a 100-inch telescope for certain types of observations. The second pair of photographs on the facing page, showing the galaxy M82, provides an example. The one on the left was taken on colour film with a 20-inch telescope, and the one on the right was obtained using a CCD on a 24-inch telescope. Combining three images taken through filters produced the colour in the CCD picture.

Since a CCD image is composed of millions of electronic pixels, comparable to the dots that make up a television picture, the astronomer can select exactly how the information is to be reconstituted. The process usually takes place with the astronomer seated in front of a high-resolution colour-television monitor twiddling knobs or typing computer instructions. The displayed image is often given an array of brilliant colour, usually unrelated to the celestial object's true hue. The result, sometimes stunningly beautiful, tells a story of intensities. Each colour represents a different intensity of light energy received by the CCD. An example of this process is the bottom pair of images on the facing page, taken by the electronic camera on Voyager 1, that reveal detail in the rings and clouds of Saturn.

When applied to a galaxy, the false-colour enhancement often produces a ragged-edged rainbow, or bull's-eye, effect around the bright nucleus. A black-and-white photograph, or even one in colour, cannot display either the minor intensity differences or the extremely faint details that become strikingly apparent in the false-colour CCD images. However, nothing touches photographic film for representing cosmic objects as they would appear to the human eye. Furthermore, photographs can be reprocessed electronically to yield some of the advantages of false-colour imagery. The two images of the full moon, **right**, demonstrate this technique. Details only subtly seen—or invisible—are unmistakable in the false-colour image derived from a normal colour photograph. The same method was used for the pair of images of the Whirlpool Galaxy, M51, at the top of the facing page. Finally, the technique can be extended one step further

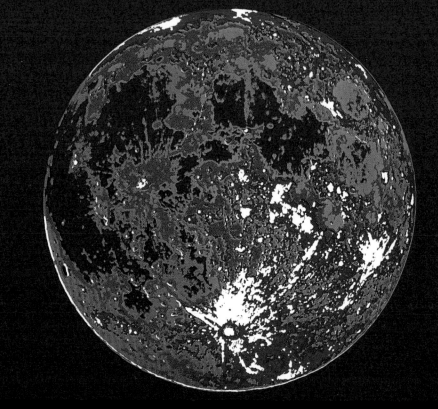

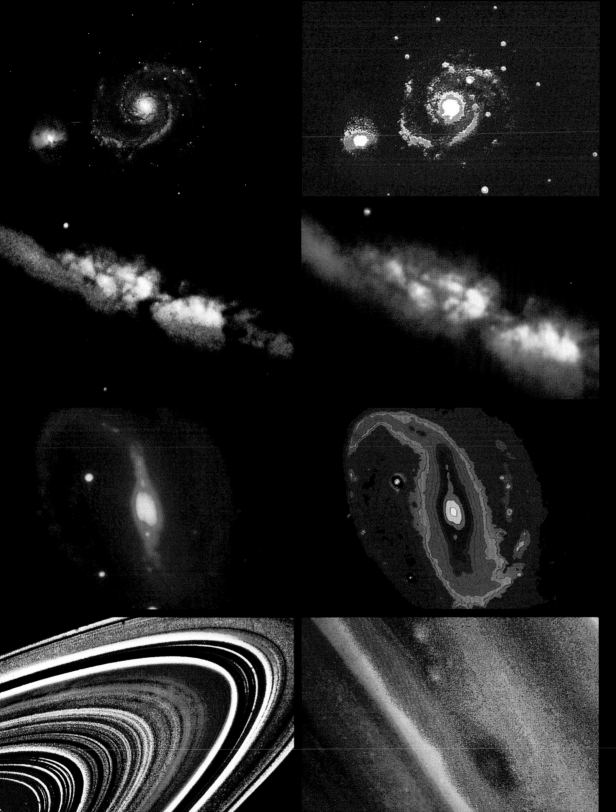

(third pair of images, this page), by taking a true-colour CCD image and enhancing it with false colour. By the end of the century, the vast majority of celestial images obtained for research purposes will be electronic, rather than photographic.

emits vast amounts of high-energy radiation, usually in the form of gamma rays.

The first hint that the galactic core consists of one or more giant black holes came in 1977 when high-altitude balloons fitted with gamma-ray detectors recorded intense gamma radiation coming from the direction of the galaxy's nucleus—exactly the type that theorists say could be generated only at the superheated inner fringes of an accretion disc spinning around a gigantic black hole. The accretion-disc model also fits with findings of radio astronomers who mapped radiation from the galactic nucleus (radio waves can penetrate the gas and dust that block visual light from the nucleus).

Occasionally, a large amount of material, perhaps an entire star, plunges toward the hole, causing a burst of radiation, like throwing gasoline on a fire. The more distant material falling toward the hole is then blasted out into the galaxy. Indeed, radio-astronomy teams have detected doughnuts of material ejected from the galaxy's nucleus like smoke rings, which have been attributed to periodic eruptions. Every few thousand years, the galaxy probably hiccoughs in this way.

Quasars: Light Fantastic

When quasars were discovered in 1963, they instantly became astronomy's number-one enigma. Measurements of their spectral redshifts, caused by the expansion of the universe, indicated that quasars were at enormous distances from Earth, close to the fringe of the known universe. Yet a quasar's radiation is emitted from a zone barely larger than 10 times the diameter of the Earth's orbit around the sun. From such a relatively minuscule area, the quasar pumps out energy equivalent to that radiated by trillions of stars like the sun.

There were no theories at the time to explain how so much energy could be coming from such a compact source. Astronomers wondered whether quasars might be less powerful than they seem. Perhaps they were closer than first thought, somehow masquerading as remote beacons. However, in the last few years, the evidence has become overwhelming that not only are quasars out among the distant galaxies, they *are* galaxies—more specifically, the nuclei of galaxies in a particularly violent stage of evolution. Due to their exceptional radiance, quasars can be seen at enormous distances, whereas the galaxies they inhabit are, by compari-

son, almost invisible. But three pieces of crucial evidence have virtually clinched the case.

New electronic detectors far more sensitive than photographic film have been united with the world's largest telescopes. The powerful combination has revealed previously unseen groups of faint, but otherwise normal, galaxies surrounding several quasars. Astronomers realized that quasars, if randomly scattered across the universe, would probably not appear among groups of galaxies. A few might appear by accident, but so many quasars have been noted in such positions that there has to be a connection. The most reasonable assumption is that quasars are the brightest members of the galaxy clusters.

Astronomers examining the dim haloes of light around most quasars have shown that the quasar "fuzz," as it is called, is the combined light of billions of stars similar to those found in galaxies such as the Milky Way. They conclude that quasars are rare superbrilliant cores of galaxies. Further proof comes from a few images that reveal hints of spiral arms emerging from some quasar nuclei.

Perhaps the most persuasive piece of evidence linking galaxies and quasars is a supernova discovered in 1984 in the luminous fuzz surrounding quasar QSO 159 + 730, about 1.5 billion light-years away. This is the first time a supernova has been found in a galaxy in which a quasar resides, although supernovas have been observed in normal galaxies hundreds of times during the last century. Supernovas can be used as distance gauges. Like streetlights, supernovas have approximately the same maximum luminosity. The dimmer they appear, the farther away they are. Judging from its brightness, the supernova seen in the quasar fuzz of QSO 159 + 730 is one to two billion light-years away, corresponding to the distance determined from the redshift. Astronomers at last had a completely independent method with which to confirm a quasar's distance.

But identifying where quasars are does not establish what they are. Within a zone barely larger than the width of Pluto's orbit, a typical quasar generates energy equivalent to that of hundreds of galaxies like the Milky Way. The compacted focus of the quasar's titanic energy release cannot be explained by ordinary stellar radiation. No matter how powerful or massive the stars or how densely they are concentrated in the galactic core, they simply do not have the brute output of a quasar.

This six-step zoom into the heart of a quasar begins at upper left within the confines of a galaxy cluster dominated by a giant elliptical galaxy. Jets emerging from the galaxy are seen in most quasars but are shown here in more detail. The frame below reveals some of the thousands of globular clusters surrounding the giant elliptical. Unseen is a halo of gas that is trapping a spiral galaxy which has ventured too near the massive elliptical. Sputters punctuating the jets indicate periods of enhanced activity at the core prior to their ejection. In the next increment, the core of the huge galaxy is exposed. Here, five black holes rip stars into light-year-long tendrils of gas that are ultimately funnelled into the holes. The dominant hole has the mass of a billion suns. The fourth frame, upper right, closer to the galactic core, shows the monstrous gravity whirlpool 100 times the diameter of the solar system. The penultimate scene reveals the source of the quasar's jets, and the final view takes us to within a few solar-system diameters of the quasar engine itself. The powerhouse black hole, largely concealed, is in the doughnut hole at centre. The doughnut is the hole's accretion disc.

= 1,000,000 light-years

= 1 light-month

= 100,000 light-years

= 1 light-week

= 1 light-year

= diameter of solar system

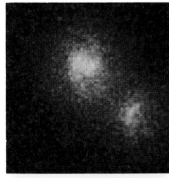

Missing most of their spiral arms, galaxies float like minions waiting to serve the two giant elliptical galaxies in this view, **right**, *of the central zone of the Virgo galaxy cluster. Close encounters with the giants have already cost the spirals their arms. Some of them may eventually be swallowed entirely by the gravitationally dominant ellipticals. This region marks the core of the Virgo super-cluster, our local supercluster of galaxies. The two Hubble Space Telescope images,* **above**, *reveal some of the most distant galaxies yet seen as anything more than mere points of light. At least five billion light-years away, these galaxies are probably in the process of merging. Many similar Hubble images suggest that galaxy interactions and mergers must have been com-monplace billions of years ago.*

Only one known object does: a giant black hole.

A massive black hole has the potential to gener-ate radiation 100 times more efficiently than the thermonuclear fires at the core of a star like the sun. A black hole's power plant is fuelled by anything that is sucked into the vortex. The enormous grav-itational whirlpool twirls infalling matter into an accretion disc and accelerates it to nearly the speed of light. The process generates prodigious quantities of energy that burst into space as in-tense gamma and x-ray radiation just before the doomed matter plunges into the hole. The key to a quasar's fountain of radiation seems to be an abundant supply of matter for its black hole.

Stars and gas clouds swarm at the centre of all galaxies, but to yield the kind of output emerging from quasar nuclei, the equivalent of several hun-dred thousand Earths—or one whole star—in dust and gas would have to be consumed annually. Quasar black holes are feasting on an apparent banquet of infalling material not available in ordi-nary galaxies. However, they would not swallow whole stars. A doomed star would never make it to the hole's event horizon in one piece. It would be-come entangled in the vortex of the accretion disc and shredded, or "spaghettified" as one astrono-mer put it. The huge amount of matter swirling into a quasar black hole (a billion trillion tons each second) would form a blazing doughnutlike whirl-pool around the hole that would churn out more en-ergy than the radiation from trillions of stars. The radiation blast wave itself would likely boil any stars that ventured near the hole. The resulting shreds of stars and tendrils of gas swirling into the hole would be incandescent, creating a terrifying scene of raw power unmatched in the universe.

Black holes with abundant food supplies are messy eaters. They grab more than they can swal-low. The surplus seems to be ejected in two jets perpendicular to the accretion disc, presumably over the hole's rotational poles. Why the hole fails to suck in everything that approaches it is unclear,

Fountaining a geyser of plasma and gas, a five-billion-solar-mass black hole at the nucleus of a giant elliptical galaxy is seen from the surface of a hypothetical planet. Stars near the hole are turned into comet-like bodies as they head to their doom, eventually merging with the black hole's huge accretion disc.

but the dual-jet phenomenon is seen in many galaxies with active nuclei as well as quasars. In some cases, huge clouds of matter have been hurled millions of light-years, more than the distance from Earth to the Andromeda Galaxy.

It now seems likely that almost every galaxy has a central black hole created during galactic birth. An initial intense period of star formation and massive-star supernovas at the crowded galactic nucleus would have left many black holes in the 10-solar-mass range. These would have grown by colliding and merging and by feeding on the nebulas that were plentiful then. The larger of two colliding black holes would emerge with the added mass of the consumed hole. Eventually, one dominant hole would define the galactic nucleus. But a normal spiral galaxy's nuclear zone—as evidenced by our galaxy—does not create a billion-solar-mass black hole. Something else is required.

Just a tiny fraction of galaxies are quasars. The nearest one is more than a billion light-years away.

Obviously, a combination of rare circumstances is needed to provide enough material for a monster galactic black hole to pump out quasar-level radiation. The creation of giant elliptical galaxies through galactic collisions may be the only phenomenon violent enough to do it. Clouds of dust and gas in the interacting galaxies would collide, lose momentum and plunge into the galactic nucleus during such encounters. Consuming matter at the rate of up to a million Earth masses per year would soon inflate the galaxy's primordial black hole to millions and possibly billions of times the mass of the sun. A one-billion-solar-mass black hole would be about the diameter of Pluto's orbit, corresponding to the apparently small celestial nozzle from which the quasar energy is pumped. The larger the black hole, the more energy can be released—provided it has copious amounts of fuel.

Quasars do not last forever. The engine for quasar fireworks requires a hole of at least several hundred million solar masses that functions at full

101

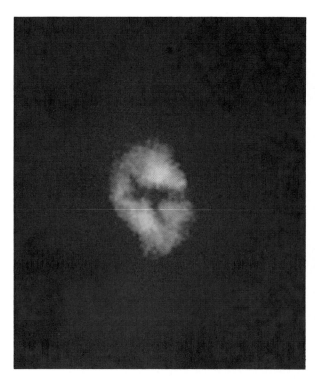

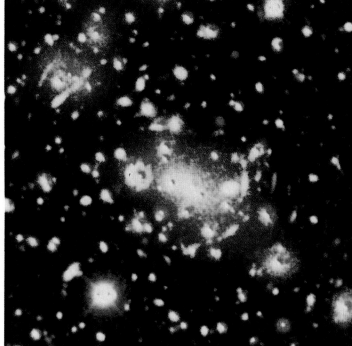

Four modern views of galaxi *(all obtained in 1991 or 1992) reveal detail never seen before. A distinct X marks the core of the spiral galaxy M51,* **far left**, *also known as the Whirlpool Galaxy, about 25 million light-years away. The wider bar is thought to be a colossal disc of gas and stellar debris more than 100 light-years across, seen edge-on and whirling around a black hole of perhaps one million solar masses. This concept is depicted in the illustration on page 85. The other bar in the X may be a second black hole accretion disc at an angle to the first. Probing deeper into space,* **centre left**, *the 158-inch telescope at Kitt Peak was used to image the cluster of galaxies A2218, about two billion light-years away. The small bluish arcs represent the light from more remote galaxies warped by the gravity of the foreground cluster. The warping is far greater than can be accounted for by the visible galaxies alone. Vast amounts of dark, invisible matter must exist in the cluster—another example of the stuff that has eluded identification for more than two decades. Evidence for a three-million-solar-mass black hole is seen in this Hubble Space Telescope view,* **centre right**, *of the condensation at the core of M32, one of the Andromeda Galaxy's elliptical companions. Stars are just light-weeks apart in this region, where 100 million stars are squeezed into 400 cubic light-years of space—the same volume that typically contains only one star in our spiral arm of the Milky Way. Graininess in this image is the brightest of the swarms of individual stars. Finally,* **far**

102 **right**, *the Hubble Space Tele-*

throttle in bursts lasting about a million years. At its peak activity, the black hole creates such an enormous energy flow that nearby gas and dust clouds are blown back, reducing the fuel supply. The action then subsides for a few million years, the displaced material gradually returns, and the cycle repeats until the hole depletes the galactic core zone to the extent that maximum consumption is never resumed. It is doubtful whether a galaxy can remain in a quasar phase for more than two billion years. Thus many galaxies probably harbour giant central black holes that may have the potential to be quasars but are currently inactive.

The most remote quasar is approximately 12 billion light-years from Earth. Far more quasars are seen at great distances than random distribution would dictate. Astronomers conclude that quasars were abundant during the first few billion years of the universe's existence, which is what we would expect if quasars are produced as a result of galactic collisions, because such collisions were more common then than they are today.

Emperor Galaxies

Galaxies are often grouped in families with dozens or hundreds of members. Dominating the central regions of these galaxy clusters are obese giant elliptical galaxies, ranging up to, and perhaps even exceeding, 50 trillion times the mass of the sun. The larger giant ellipticals are known in astronomical jargon as cD galaxies. These are the masters of the universe and the bullies of the galaxy families. More massive than any other discrete stellar structure, they are the gravitational hubs of galaxy clusters. Nearby galaxies are trapped, looping in vast orbits around the cD galaxies.

These emperor galaxies may be cannibals, growing fat at the expense of other galaxies that are partially denuded or completely torn apart and swallowed up as they pass the giant ellipticals. Sometimes, more than 100 smaller galaxies swarm around a cD, many of them barely more than the nuclei of once elegant spirals. Cannibal cD galaxies look like vastly overgrown globular clusters in photographs, but other instrumentation reveals that they are surrounded by enormous haloes of stars and gas evidently ripped from passing galaxies. The halo acts as an effective net for ensnaring gas from more distant galaxies that pass by. The cD galaxy thus extends its influence and leaves its neighbour galaxies as star skeletons without nebulas for new generations of stars. Meanwhile, the trapped gas slowly falls toward the cD, adding to its mass.

Some cD galaxies and giant ellipticals have

quasarlike jets extending out from their nuclei to a distance that is several times the galaxy's diameter. There may be a direct connection between that activity and quasars. During a violent encounter, when huge quantities of gas fall into the cD from another galaxy, the big galaxy's central black hole might get stoked up to quasar levels. The jets could be evidence of such activity. The nearest cD, known as M87, has a jet and has been diagnosed as containing a central black hole of five billion solar masses, easily big enough for quasar-level bursts.

All this violence among and within galaxies is alien to our star city, the Milky Way. Our nucleus black hole is puny compared with the one in M87. And the closest we have come to a galaxy collision is the event that caused the disruption and recent burst of star formation in the Large and Small Magellanic Clouds, our satellite galaxies. Astronomers are uncertain about what happened, but some mild collision-related event probably triggered the activity so clearly visible in photographs of the Large Magellanic Cloud. Sooner or later, all satellite galaxies are destined to be either ripped apart or swallowed up or left totally devoid of gas and outrider stars. The Andromeda Galaxy's two small elliptical satellite galaxies may have suffered the last fate. However, both Andromeda's companions and our satellite galaxies are too small to do major

structural damage to the main galaxies.

The placid history of the Milky Way Galaxy seems largely due to its location in a small dispersed galaxy cluster called the Local Group, which lies on the outer fringe of the Virgo Galaxy Supercluster. (A galaxy supercluster is a collection of galaxy clusters that seem to be gravitationally bound to each other but do not merge.) The Local Group galaxies have been producing generation after generation of stars for billions of years without bothering their neighbours. By contrast, ellipticals are sometimes scenes of galactic cannibalism. Do such drastic differences between conditions in our galaxy and those in ellipticals or in disturbed galaxies in dense clusters have any bearing on the prospects for life in the universe? Galactic nuclei seem to be the places to avoid. Spiral-arm stars never venture close to the core. Our sun has a near-circular orbit around the nucleus. Elliptical galaxies are a tangle of stellar orbits. Most of their stars swing in looping paths that carry them within a few thousand light-years of the core. Just how this would affect a planet of such a star is not known. A planet like Earth would not be disturbed from its solar orbit, but the occasional radiation bath might be a problem. Probably the most we can say is that ellipticals cannot be ruled out of the life game. Even if they are, there are billions of spiral galaxies in the universe.

(continued from facing page) scope probed deep into the interior of the giant elliptical galaxy M87 to spy a dense concentration of mass (seen as a brilliant point) that astronomers interpret as a two-billion-solar-mass black hole, the largest so far identified. The galaxy's plasma jet extending to the right may be something like the jet depicted on page 101.

INTO THE ABYSS

By human standards, one cubic light-year is an enormous chunk of space, enough to hold the sun, its orbiting family of planets from Mercury to Pluto, a million roving asteroids and a trillion comets on giant looping paths which carry them so far out that mother sun is reduced to a bright star in the firmament. But on the scale of the universe, it is merely our backyard—just one infinitesimally tiny pocket of the cosmos. The total volume of the known universe is a million trillion trillion cubic light-years.

Suppose one of those million trillion trillion cubic light-years were selected totally at random as a site for exploration and a survey of surrounding space. How would the cosmic landscape vary from one site to another? Ninety-nine times out of a hundred, an intrepid explorer willing to embark on such a blind voyage would emerge in a pristine vacuum embedded in total blackness. Absolutely nothing would be visible to the unaided eye. Resorting to binoculars, the traveller might spy a few smudges of light—some remote galaxies—but statistically, that would be unlikely too.

The universe is almost entirely empty space. Planets, stars and galaxies are scattered here and there in an abyss of nothingness (at least as far as human senses are concerned). There is one galaxy for every million trillion cubic light-years of space and one star for every billion cubic light-years. But they are not evenly distributed. Stars are swarmed in galaxies, and galaxies congregate in clusters. On the largest scale, the clusters, which are separated by enormous voids, are themselves arrayed in groups known as superclusters. An observer

plunked randomly in the universe would probably land somewhere in one of the voids that occupy most of the cosmos. In regions where galaxies exist, the superclusters are like island archipelagos scattered across the cosmic ocean.

Recent surveys of galaxy distribution have uncovered a structure, perhaps a grand design, to the universe. The voids are vast, roughly spherical areas that can extend for hundreds of millions of light-years. The only substantial objects within them would be occasional errant galaxies or globular clusters that long ago escaped the family gravitational attraction of the galaxy superclusters.

Around the "surfaces" of the voids lie the galaxy superclusters. The entire affair resembles the cellular structure of soap bubbles: where two voids meet, a sheet of galaxies is likely; zones of multiple intersections seem to produce denser ribbons and tendrils of galaxies. The junctions of several void surfaces at one site can generate the most populous knots, marking the cores of galaxy superclusters.

Just as a sink full of soap bubbles is mostly air with a little soap, so the universe on the whole is primarily vacant. However, layer upon layer of bubbles give the impression of substance, as do photographs crowded with galaxies.

The universe's cellular anatomy is the most recent link in a chain of cosmological revelations that date back to 1924 when Edwin Hubble used the Mount Wilson 100-inch telescope to examine individual stars in the Andromeda Galaxy. This was the first conclusive proof that galaxies other than the one we inhabit exist. For decades prior to Hubble's triumph, the nature of the spiral nebulas, as galax-

Now entertain conjecture of a time / When creeping murmur and the poring dark / Fills the wide vessel of the universe

WILLIAM SHAKESPEARE
Henry V

An island of more than 200 billion stars, ponderously turning once in about 200 million years, floats in the blackness of deep space 15 million light-years from Earth. Known only by its catalogue number, M83, this galaxy is about the same size and mass as the one our sun inhabits.

105

Woven across the cosmic landscape like the threads in some vast tapestry, **right**, galaxy superclusters define the ultimate structure of the universe. This image is a plot of nearly a million individual galaxies as seen from Earth and represents about half the entire sky. Far from being scattered at random, galaxies are clustered, and the clusters congregate in superclusters that form the filaments seen here. The chain of galaxies, **lower right**, extending from upper left to lower right, is a small sector of a nearby supercluster. (The galaxies are the fuzzy ovals; stars that happen to be in the line of sight, belonging to our own galaxy, are dots.)

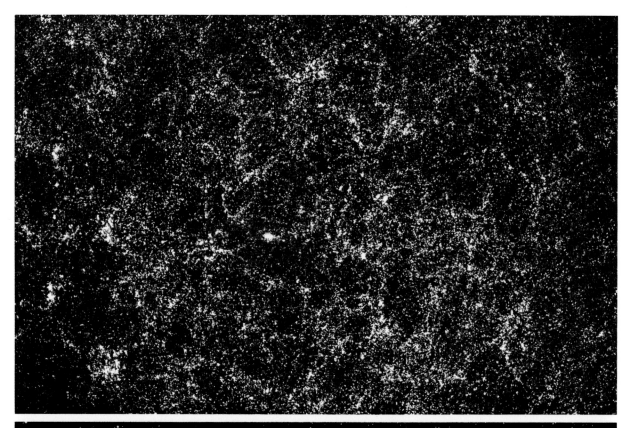

ies were then called, was uncertain. Before that, philosophers swung just as much weight as astronomers in discussions of the structure and extent of the cosmos. The discovery of the universe is almost entirely a 20th-century enterprise. Today, the uncharted territory is on more distant shores, where the fireball that gave birth to our universe may have created a multitude of other universes.

The Expanding Universe

In less than a minute, the universe will increase its volume by a trillion cubic light-years. Propelled by the force of its explosive birth 15 billion years ago, the universe is expanding like an inflating balloon.

Albert Einstein mathematically predicted the expansion as a side effect of his general relativity theory in 1916, but the idea was so revolutionary at the time that even Einstein himself balked at the implication of his own equations. In a move that he later called "the biggest mistake of my life," he inserted a term in the key equation to keep the universe static. Meanwhile, the first evidence that the galaxies are, in fact, receding from one another was being gathered by astronomer Vesto Slipher at Lowell Observatory in Flagstaff, Arizona, using the same telescope that had been built to study the canals of Mars. The canals proved to be optical illusions, but the receding galaxies are fact.

Slipher announced his findings before the exact nature of galaxies was understood, so it was not clear at the time exactly what these objects speeding away from Earth were. But there was no doubt about the fact itself. It hinged on analyses of the spectral "fingerprints" contained in the galaxy light. Those fingerprints are the positions of lines in the spectrum of starlight, caused by glowing gases such as hydrogen or the absorption of starlight by gases. The lines are always in the same location relative to one another, like the keys on a piano. But shifts in the entire set of lines reveal whether the galaxy that contains those stars is approaching or receding. The amount these lines are shifted away from the positions they occupy in the spectrum of a stationary laboratory source indicates a velocity in the observed object. It is like having two piano keyboards, one in concert pitch, the other, with an identical-looking keyboard, transposed a tone and a half lower so that the pianist has to play two keys to the right.

If the lines are shifted to the red, or longer, wave-length end of the spectrum, the source object is moving away from us. A blueshift means it is moving toward us. The effect is produced by the same principle that causes the pitch of a train horn to change as the train approaches the listener and then recedes. The wavelengths of sound are compressed to a higher pitch as the train draws nearer and are stretched to a lower pitch as it speeds away. Light displays similar wavelike characteristics to an observer studying approaching and receding celestial objects.

Slipher found that the vast majority of galaxies had a redshift, indicating the wavelengths of light were being stretched to lower frequencies as the galaxies receded. Every survey of galaxies since then has confirmed that this is correct. Furthermore, these surveys have shown that the farther away a galaxy is, the faster it is receding. It would seem, on the face of it, that everything is moving away from us, placing our galaxy at the centre of the universe. Could this in fact be the case? And why do more distant galaxies move away faster?

A loaf of raisin bread provides a reasonably accurate picture of the situation. When the loaf is being prepared for baking, the raisins are randomly distributed throughout the dough. As the dough rises, it expands fairly evenly to several times its original size. During the expansion, the raisins are carried along with the dough, each moving away from the others. Any one raisin will "see" all the remaining raisins receding from it, which creates the illusion that each raisin is at the centre of the expansion. As the loaf doubles in size, a raisin one inch away will recede to two inches, while one that is two inches away will have moved to four inches. The more distant raisin therefore moves twice as far in the same time interval. Thus during the expansion, a raisin twice as far away recedes twice as fast.

This analogy, in which raisins are galaxies, clearly demonstrates both why we seem to be at the centre of the expansion and why the farther away a galaxy is, the faster it seems to be receding.

The raisin-bread model portrays another aspect of the universe, in that it shows the raisins being carried by the dough, rather than moving through it. The dough represents space in the real universe. Although space is, for practical purposes, a vacuum, it is part of the universe. A point in the universe exists in both space and time whether there is any matter there or not. Light and other forms of radia-

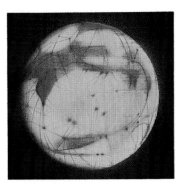

In one of the great ironies of astronomy, the telescope that was designed and built expressly to observe the canals of Mars (which proved to be nonexistent) was the instrument that revealed the first evidence that the universe is expanding—one of the most profound discoveries in the history of science.

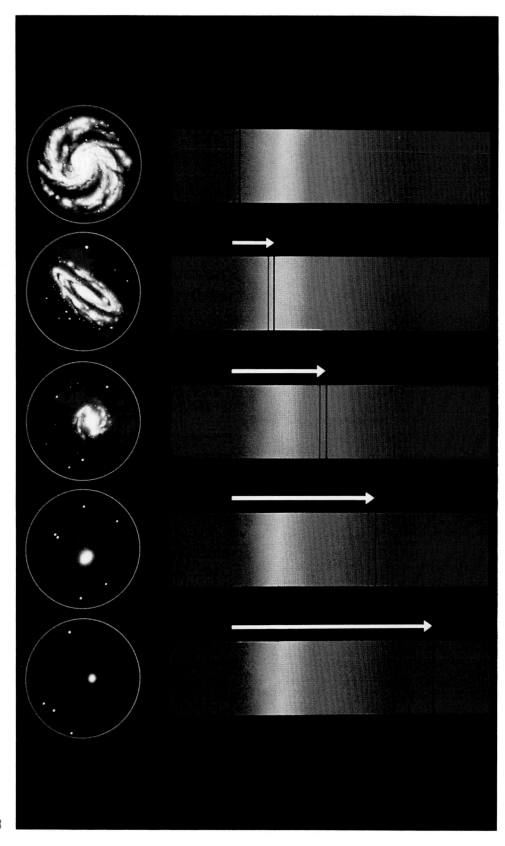

tion propagate through space, and gravity holds sway throughout the universe (although the mechanism by which its force issues through space is still unknown). It is space that is expanding, carrying the galaxies along with it. Galaxy clusters resist this trend, though; the gravitational grip of the cluster members on each other is usually strong enough to hold the group together. The superclusters, more loosely bound than the clusters they comprise, are also part of the expansion, but their constituent galaxy clusters drift apart at a slower rate than the overall expansion due to the mutual gravitational pull of the group. The full expansion rate is only observed in the redshift of galaxies in superclusters receding from our supercluster.

But a problem has developed in this neat scenario: the galaxy clusters do not seem to be massive enough to produce the gravitational pull needed to hold themselves together. To keep from flying apart, the clusters should have more or larger galaxies. Yet the clusters exist. The conundrum, intertwined with some related aspects of cosmology, has emerged as the paramount enigma in late-20th-century astronomy. The mystery began about two decades ago when astronomers analyzed the motions of galaxies in clusters by measuring the differences in the redshifts of the cluster members. That allowed the velocities of individual members to be compared with the group average. The result of the investigation revealed that member galaxies are moving around in ways which indicate the clusters are not flying apart—at least not significantly. The conclusion: some invisible stuff is holding galaxy clusters together.

The confirmation that there is more to the universe than meets the eye came soon after, when researchers estimated the total number of stars in the member galaxies of clusters (the estimates are based on the luminosity of the galaxies). There appeared to be less than one-thirtieth of the amount of material needed to keep the cluster together. Judging from the visible mass of their galaxies, the clusters should have dispersed long ago, leaving galaxies evenly spread across today's sky.

The quest then became one of identifying what the invisible mass is or, failing that, where it is located. A lot of it is inside and around the galaxies themselves. Galaxies may be carrying up to 10 times more invisible mass than visible. Something other than ordinary stars is there in huge amounts. This revelation came from measurements of the

spin rate of galaxies. Stars in the outer region of a galaxy orbit much faster than traditional estimates of the galactic mass would indicate. For example, the sun takes 200 million years to complete one orbit around the galaxy's nucleus. The orbital speed is determined by the amount of material between us and the centre of the galaxy, which is about 200 billion times the mass of the sun—not much different from mass estimates used for many years for the entire galaxy. The material outside the sun's orbit has little effect on the calculation. But a star at twice the sun's distance from the centre of the galaxy "feels" the gravitational influence of all the matter—stars, planets, dust and gas—between it and the galaxy's nucleus, and its orbital velocity is determined by that total mass. Globular clusters, some of which are even farther out, feel the mass contained in a larger volume.

The motions of outlying stars and globular clusters in our galaxy and others were not known until recently, but according to new data, the Milky Way Galaxy may be 10 times as massive as the light coming from its stars indicates. Rotation measures of other galaxies revealed the same thing, with most of the invisible mass in the outer galactic fringes, though it probably exists almost everywhere. Current estimates suggest that half of the mass in our own spiral arm is invisible and unexplained.

The Andromeda Galaxy is a typical example of the enigma this presents. The light from all of its stars is 25 billion times brighter than the sun. Since the average star is substantially dimmer than the sun, this figure translates into a total stellar mass of about 300 billion times the sun's mass. And yet Andromeda's outlying objects are moving at velocities that indicate the galaxy's mass is more like several trillion solar masses. In the case of our own Milky Way Galaxy, estimates have ranged as high as two trillion solar masses. Almost every galaxy that has been investigated is similarly overweight. Yet the extra mass cannot be seen.

There are only a limited number of conventional explanations. Stars produce light, and the light can be measured, so there cannot be more normal stars than we see. Then there are the unusual stars: stellar cadavers such as black holes, neutron stars, brown dwarfs and white dwarfs, none of which give off much light. However, a census of the region within a few dozen light-years of the sun has not turned up even a small fraction of the number of white dwarfs that would be required (they would

have to be far more numerous than normal stars to account for the invisible mass). Black holes, neutron stars and brown dwarfs could more easily remain undetected and should not be ruled out, although there is no compelling evidence to support the idea that they are present in huge numbers.

Could the invisible material be compressed into a supermassive black hole at the galactic core? Such objects almost certainly exist, but hundreds of billions of solar masses would be necessary right in our own galaxy's core. Most of the galactic mass cannot be a point source because the orbital velocities of the Milky Way Galaxy's stars indicate that the mass is spread out. Also, the gravity forces from such an enormous object would be readily disclosed by the orbital motions of stars and gas clouds near the centre of the galaxy. Those motions reveal a central mass about four million times the sun's mass, nowhere near the required amount.

Huge numbers of pint-sized bodies—planets, asteroids, comets—scattered like trillions of motes of dust throughout our galaxy and others have been proposed. But modern theories of star formation cannot account for the vast number of these objects necessary to make up our galaxy's invisible deficit (for example, a million bodies the mass of Earth for *each* star in the galaxy would barely be enough). Theoretically, the creation of so much debris is unlikely without stars emerging with it. Although it cannot be ruled out completely, the idea has not gained much support among researchers.

The only definite dent in the invisible-mass enigma was the x-ray-satellite discovery of a thin, hot gas between galaxies in large, dense clusters. The gas may be thin—a near vacuum by human standards—but there is so much of it that it doubles the cluster's mass. Still, the invisible stuff in and immediately around galaxies remains a mystery. Theorists have resorted to invoking the existence of undetected but theoretically plausible subatomic particles such as massive neutrinos, axions, gravitinos, photinos and heavy leptons. The situation may in some ways parallel human perceptions of the air around us. Wind forces reveal the air's existence, but most of its gases are invisible. Only water vapour, which constitutes a tiny percentage of the Earth's atmosphere, is readily visible (fog, haze, rain, snow, et cetera). Perhaps familiar forms of matter such as stars, planets, gas and dust are a minority component of the universe. Exotic particles, totally remote from everyday experience,

There is more to galaxies than meets the eye, as revealed by this false-colour image that shows a halo of matter around the galaxy M104, not seen in normal photographs. Galaxies contain up to 10 times more mass—most of it invisible—than their visible stars indicate. The nature of the invisible mass remains a mystery. However, distances to galaxies, **facing page**, *can be gauged in a straightforward manner: the farther, the faster. That is the effect of the expanding universe. The measuring tool is the spectrograph that captures the telltale absorption (dark) lines in a spectrum of a galaxy. The farther the lines are displaced to the red—redshift— the faster the galaxy is receding and the more remote it must be.* **Following page:** *The Milky Way Galaxy seen from above the clouds of a hypothetical planet in the Small Magellanic Cloud. At left are the stars and nebulas of the Large Magellanic Cloud, a satellite of the Milky Way and our nearest galactic neighbour.*

109

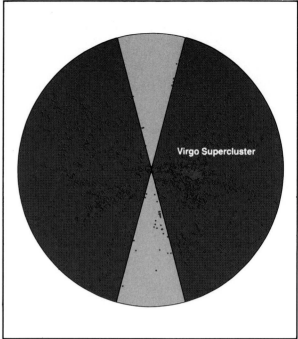

Virgo Supercluster

Two giant elliptical galaxies dominate this galaxy cluster in the constellation Centaurus, **right**. *They almost certainly have ingested some nearby galaxies and have already stripped most of them of their spiral arms. Dense galaxy clusters like this often are the hub of galaxy superclusters—clusters of clusters. The Milky Way Galaxy and the Local Group (our neighbourhood galaxy cluster) are at the outer edge of the Virgo supercluster,* **far right**. *The diagram is a plot of all known galaxies out to a distance of 150 million light-years. Each dot is a galaxy. The Milky Way is at the centre. The grey wedges are zones of the sky blocked from view by dust and gas in the plane of our galaxy. The Virgo supercluster's hub is marked by the Virgo galaxy cluster, pictured on page 100.* **Facing page:** *From our viewpoint in the Milky Way Galaxy, we seem to be at the centre of the universe, but that is an illusion. We see the universe as it once was, not as it is today. Only events in our immediate vicinity are happening now; the rest is a ghostly relic of the past. The majority of quasars we see are 5 to 12 billion light-years into space, occurrences from a long-ago era when galaxies were closer together and galactic collisions more frequent. Reaching still farther into space and deeper into the past, the cosmic background radiation, the strongest evidence of the Big Bang, is seen in every direction. To our perception, the genesis explosion is not a point in space but is everywhere. It defines the edge of space. No matter which galaxy is selected as a viewpoint, the universe would*

112 *look the same.*

may actually be the universe's major ingredient by mass. That idea meshes with developments in another area of cosmology: the universe's creation.

The Big Bang

Soon after astronomers digested the stunning fact that the galaxies are speeding away from each other, they faced an even more startling conclusion: the universe must have had a beginning. The expansion, it was reasoned, was generated by a monstrous explosion—prosaically dubbed the Big Bang—whose multi-billion-degree fury created the fabric of expanding space that hurled matter in all directions. The effects of the genesis explosion can be seen today as the galaxy superclusters dash away from each other. But, some astronomers asked, could it be an illusion? They wanted proof that the Big Bang happened. The debate peaked in the 1950s with the introduction of the steady state theory, which suggested that the expansion is caused by the *continual* creation of matter and that there was no beginning, nor will there be an end.

The question can be probed because on the scale of the universe, distance is a time machine. The more distant something is, the longer its radiation takes to get here and the younger it appears. Thus the explosion that preceded the origin of the galaxies should be detectable if astronomers can

look deep enough. It would appear in all directions because the space-time fabric emerged from it, so any sight line would lead to the creation event. Princeton physicist Robert Dicke laboured for years to determine exactly what the Big Bang would look like. He calculated that we might still detect the embers of the creation blast vastly redshifted by the fact that our galaxy is racing away from the site of its birth at almost the speed of light. Since temperature is related to wavelength, Dicke said, the redshift would be so great that the radiation would not be visible light but, rather, very cool and invisible microwave radiation. This is comparable with the train in our redshift analogy receding at such speed that the sound of its horn is below audible levels.

Unaware of Dicke's work, Arno Penzias and Robert Wilson of Bell Laboratories in Holmdel, New Jersey, were testing a new microwave antenna in the spring of 1965 when they noticed a faint but persistent signal coming from all directions. It was exactly what Dicke had predicted. It is now called the three-degree background radiation because that is how cool the redshifted creation fires appear (three degrees Celsius above absolute zero). The Big Bang happened. Its echo can still be heard 15 billion years later. For their serendipitous find, Penzias and Wilson later collected a Nobel prize.

With the reality of the Big Bang firmly established, attention has now focused on the universe's age and ultimate destiny. Age estimates are keyed to galaxy-cluster distance estimates. If the clusters are as close as some researchers contend, the universe is about 10 billion years old. If they are as remote as other equally respected investigators say, we live in a 20-billion-year-old universe. Despite major efforts by hundreds of astronomers in the past few decades, the range of estimates has not been whittled down. The age used in this book —15 billion years—is simply a median.

As for the destiny question, there are two possibilities: either the universe continues its expansion forever or the outrush will stop and the whole affair will collapse again. Because of the space-time geometry involved in the two scenarios, astronomers call the ever-expanding case an open universe and the collapsing option a closed universe. During the 1970s, the tide of evidence heavily favoured continual expansion. But the pendulum has swung back in the 1980s, and today, it hovers about midway between those views.

Which of the two theories is the more plausible is contingent on how much mass there is in the universe. If the universe exceeds the critical mass, it will ultimately yank itself back into its primordial egg. Just as a rocket that does not have the thrust to reach escape velocity will plunge back to Earth, the galaxies may be pulled back to their origin if the universe as a whole has sufficient mass to force complete deceleration of its parts. Depending on whose theoretical models are selected, the mass required to do this varies somewhat. Even so, the total mass of all the galaxies (including their invisible but massive haloes and the enigmatic phantom mass in galaxy clusters) is between 10 and 30 percent of that needed to reverse the expansion. If that is all there is, the universe is open and ever expanding. However, many theorists have suggested the rest may be there as well, disguised in some way we have yet to recognize.

There is also a philosophical component in the equation. Some astronomers find an open universe untidy, even unpalatable. An ever-expanding universe is dynamic and evolving as we see it today. But ultimately, it becomes nothing more than a hearse transporting the exhausted corpses of black galaxies ever outward into a darkened eternity. Conversely, there is something reassuring about a cosmos that has a life span, will end in the same manner that it began and will perhaps rebound into a fresh universe and begin the cycle anew.

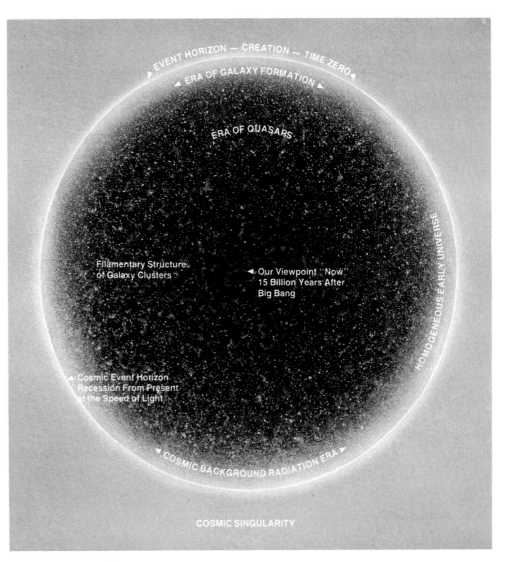

Cosmic Inflation

Until recently, it was assumed that whatever caused the Big Bang would remain forever inscrutable. But a new wave of theoretical studies may have placed the genesis scenario within reach. The breakthrough occurred in the early 1980s when investigations of the interaction of the smallest subdivisions of matter, a field known as quantum physics, began to have explicit applications to astronomy—a union that has apparently yielded a *reason* for the explosive origin of all that we know. Previous theories explained the existence of our expanding universe by citing the Big Bang, but none plausibly accounted for the Big Bang's original cause.

Quantum physics predicted that a fraction of a second after the genesis instant, the universe was

113

in an unimaginably compressed state, confined to a volume the size of a baseball. An object the mass of our sun would have been the size of an atom, and something as dense as lead would be, comparatively, the thinnest of gases. It was a seething cauldron of quarks, leptons and gluons, at a temperature 100 billion trillion times higher than inside an exploding thermonuclear bomb. From this primitive state, all matter and energy we perceive today emerged. But that is where the conjecture stopped. Answers to questions addressing just where the ingredients for the primordial energy ball came from seemed totally inaccessible.

However, new developments in quantum physics make it possible, in theory, to trace events back to a point only one trillion trillion trillionth of a second after the creation, when the universe was smaller than a proton. Temperatures and pressures were so prodigious that the four fundamental forces of nature—the electromagnetic force, the weak and the strong nuclear forces and gravity—must have been unified with equal and indistinguishable strengths. It was a time of supreme elegance and ultimate simplicity. As the embryonic universe expanded, the temperature dropped, permitting gravity to release itself from the other three forces. The expansion continued for another billionth of a trillionth of a trillionth of a second until the next transition occurred: the separation of the strong nuclear force.

That event, known to physicists as symmetry breaking, is critical to the new theories. It can be thought of as a change of state, something like the boiling-point transition from liquid water to gaseous water vapour. Although the water changes little from a temperature of 212 degrees F, there is a huge expansion in volume when it becomes a gas. Similarly, the energy of symmetry breaking in the primal universe resulted in an instantaneous injection of a vast amount of energy that drastically enlarged the universe. The primordial cosmos "hyperinflated" to at least a billion trillion times its former size—a greater increase than a grain of sand expanding to the size of Earth.

The new-inflationary-universe scenario includes the astounding prediction (still controversial) that within the supercooled, hyperinflated cosmic crucible, a nearly infinite number of universes bubbled into existence like foam, stabilizing the structure. Our baseball-sized universe, whose origin seemed so mysterious a few years ago, was one of

them. From that point until today, about 15 billion years, each of those universes has been evolving—ours among them. The expanding universe of stars and galaxies that we inhabit may be but a tiny segment of the whole of creation, an atomlike blip in the body of a colossal cosmic realm.

Some of these universes could be similar to our own; others might be composed of antimatter (where positive and negative charges of subatomic particles are opposite to our universe). Still others may be ruled by laws with fundamental constants differing from those that govern the forces of nature in our universe. For example, gravity could be weaker or stronger relative to the other forces. A universe under these alien conditions might disperse its matter so rapidly that stars and planets could not form. Conversely, a slowly expanding universe would have collapsed back on itself long ago, producing a maelstrom of massive black holes. (Although inflation seems to defy general relativity theory by exceeding the velocity of light, it does not. The part of the expanding volume that became our universe never broke the speed limit. Because each universe is disconnected from all others, relativity is not violated.)

Earlier, when we thought our universe was all there was, it was impossible to explain why it has the properties it does—properties that allow stars and galaxies to form and matter to arrange itself into planets and life. We were asking why the Big Bang made our universe the way it is. Now, with practically an infinite number of universes emerging from the inflationary theories, there can be a universe with every combination of properties imaginable. A universe like ours is but one of them. Just as Earth is teeming with life while the vast majority of worlds are thought to be unsuited for the emergence of biologies, our universe has stars, galaxies, planets and a unique balance of physical properties that are unlikely to exist in many other universes.

New attempts to explain the earliest moments of creation in a scientifically reasonable fashion are being taken very seriously because they predict a universe that can evolve naturally into the one we see, without some of the loose ends associated with previous theories. For example, several of the inflationary theories (variations of the same theory have been proposed) predict: (a) the amount of mass in the universe is *exactly* at the border line between open and closed—a flat universe that slows down but never quite reverses; (b) the back-

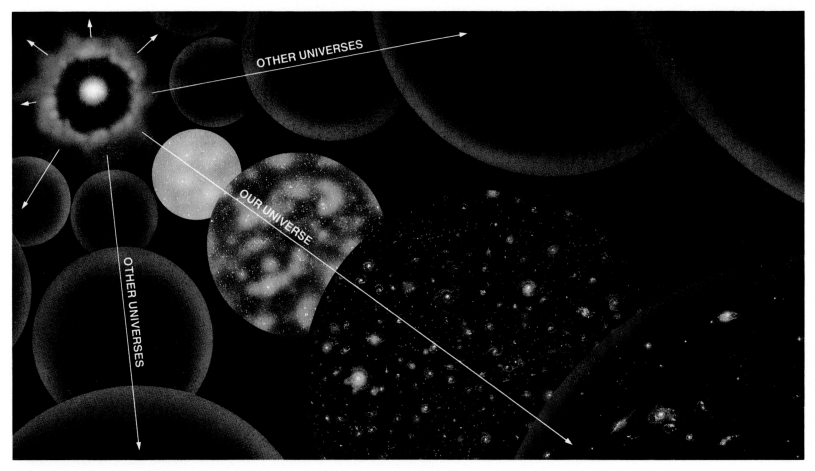

OTHER UNIVERSES

OUR UNIVERSE

OTHER UNIVERSES

ground radiation is uniform in intensity; (c) the creation event is a natural extension of known quantum physics principles rather than some miraculous occurrence. None of these conditions, which were major enigmas for decades, were accommodated prior to the inflationary scenarios. The inflationary theories predict that the invisible mass *should* be there, the background radiation *should* look the way it does, and the universe (or at least some universe) *should* exist.

Furthermore, it happens so simply. Upon hearing the concept for the first time, one physicist declared, "It has to be right; it's too beautiful not to be." Here are the essential elements:

Scientists have known for some time that subatomic particles—virtual particles—materialize from tiny fluctuations in a vacuum, then vanish. Although the process normally occurs in less than a trillionth of a second, it has been observed and can be accounted for in quantum physics theory. Since the appearance of virtual particles is a random process, it is conceivable, given an infinite amount

of time, that one of these vacuum fluctuations with exactly zero energy would occur. Such a fluctuation would not vanish after the normal ephemeral lifetime of a virtual particle. Instead, it would absorb almost unlimited energy from the vacuum, as if a hole had been punched in a dam, and emerge as a colossal fireball that would evolve just as the new-inflationary-universe theories predict.

The entire cosmos is, in essence, a reexpression of sheer nothingness, a random but predictable manifestation of the primal vacuum that predates our own universe.

The new concepts have a further appeal. They are an extension of a trend that began in the 16th century when Copernicus suggested that the sun, not Earth, is the centre of the cosmos. Since then, our sun has proved to be just one among hundreds of billions of stars in the Milky Way. In turn, our galaxy is one among billions in the observable universe. Now, our universe is seen as merely one among a multitude of cosmic bubbles drifting in a far greater ocean, the universe of universes.

A NIGHT AT MAUNA KEA

To astronomers, it is mecca: a mountaintop so high it impales the clouds, revealing by night a glittering tapestry of stars undimmed by pollution, dust or a single 60-watt porch lamp. This is Hawaii's Mauna Kea, an alien and forbidding windswept realm more like Antarctica than the tropical paradise depicted in Hawaiian vacation brochures. But despite the rigours of the high-altitude environment, the lure of pristine skies has astronomers booking time up to a year in advance for the privilege of using one of an international array of superb telescopes that dot the desolate peak.

By the early 1990s, the planet's largest telescopes will be located on Mauna Kea. Until then, the site's most advanced instrument is a 142-inch telescope designed, built and operated by Canada's National Research Council in a cooperative venture with the government of France and the University of Hawaii. Although there are six larger telescopes in operation around the globe, none is as favourably situated. Veterans of decades of research at major installations say that the Canada-France-Hawaii Telescope (CFHT), perched above nearly half of the atmosphere, may be the most powerful telescope on Earth.

The outstanding performance of the CFHT and the other instruments that share the summit is due to a combination of the 13,800-foot altitude and the generally stable Pacific air masses. Astronomers at Mauna Kea look down on the weather. The barometer at the summit barely moves from one week to the next. Only one night in six, on average, do clouds struggle up to the peak. But when they do, that nearly stratospheric realm can be slashed by instant blizzards, ice storms and hurricane-force winds. Foot-long icicles jutting out horizontally frequently decorate the railing of the catwalk around the telescope dome. Temperatures dip to the freezing point or lower almost every night. Although ever watchful for the instant storms, astronomers are delighted with the consistently clear, cold air.

Tonight, University of Toronto astronomer Barry Madore takes his turn at the big telescope. Like most modern-day astronomers, Madore actually uses a telescope only a few nights a year. The traditional image of an astronomer peering into the eyepiece of a telescope every clear night is ancient history. Madore knows that he will spend weeks, probably months, analyzing the photographs and measurements he will take tonight.

Hunched over the control console on the floor below the main dome, **page 119, bottom left**, Madore discusses the objects to be observed with resident telescope operator Kenneth Barton. They don parkas, climb a flight of stairs and emerge in the great dome, which is refrigerated to keep it at or below outside temperatures. Heat radiation from warm metal or equipment produces turbulence in the air that can distort delicate images of celestial objects. In the cathedral-like dome, Madore and Barton unconsciously converse in hushed tones. Above, the slice of sky visible through the dome's observing slit, **far right**, reveals the

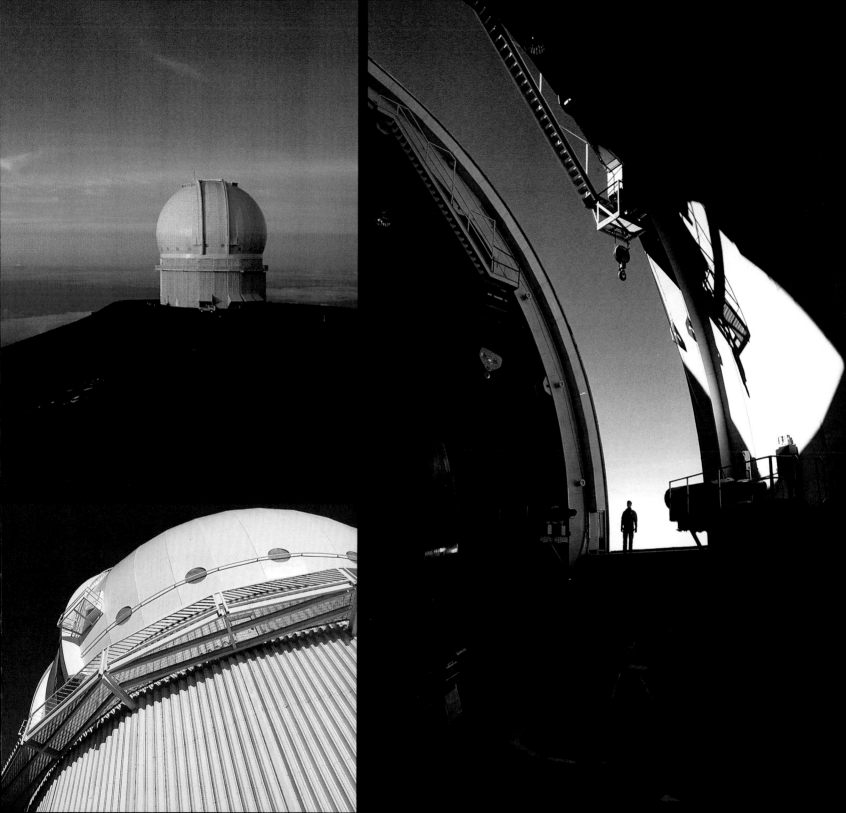

deepening indigo of twilight. Minutes later, the patch is littered with stars. It is time to begin.

Back in the control room, Barton punches computer-coded instructions on a keyboard. In response, the 300-ton telescope—resembling some colossal cannon from a Star Wars movie—wheels to vertical with the fluid grace of a delicate timepiece, **facing page, top left**. Carrying a briefcase full of unexposed photographic plates, Madore boards a tiny elevator that crawls up the curving side of the dome until it is suspended directly beside the observer's cage, a cylindrical housing not much larger than a telephone booth at the top of the telescope tube. Stepping from the elevator through a small door in the side of the cage, Madore arranges his lanky frame around a 1½-foot-diameter cylinder at the centre of the cage, which is crammed with receptacles for astronomers' research equipment. The task tonight is to photograph a relatively nearby galaxy, about 2.5 million light-years away, in more detail than has ever been possible before.

Perched atop a telescope as big as a rocket, the cramped observer's cage resembles the one-man Mercury capsules that Alan Shepard and John Glenn rode at the dawn of the space age. Madore adjusts his headset intercom and tries to get comfortable on the small wooden seat that he will occupy for most of the night. He signals to Barton that he is ready. More key punching. The telescope heels over, then stops, its 12-foot-wide mirror precisely aimed at the distant city of stars. During the one-hour exposure, the telescope's motors counteract the Earth's

rotation, keeping the galaxy in view. (The Earth's rotation produced the star trails in the time-exposure photograph, **right**. A flashlight-wielding astronomer splashed light on the dome while strolling around the catwalk.) Lying on his back in the cage, trying to ignore the cold, Madore monitors the image through a guiding eyepiece, making minor adjustments to the tele-scope's tracking motors using a push-button control paddle that he holds in his right hand.

The parallels to an astronaut in a space capsule are re-inforced as Madore relaxes for a moment to look through the hatch in the top of the cage. The stars, which he has known by name since childhood, have an unreal outer-space quality here—they rarely twinkle. Turbulence from weather fronts and atmospheric cir-culation, which causes the twinkling familiar to every casual stargazer, is almost nonexistent on Mauna Kea.

Back in the guiding eyepiece, the galaxy's twirling arms are diaphanous extensions from its hazy starlike nucleus. Light that has been on its way to Earth since well before anything human stood upright silently ends its journey in the emul-sion of Madore's photographic plate. When developed, the photograph will reveal exqui-site detail, a churning continent of a hundred billion suns.

Within a week, Madore is back in his cramped University of Toronto office, anxious to begin examining the results of his night on the mountain. He gingerly places a photographic plate of the galaxy M33 on a light table (a photograph of this galaxy appears on page 16). The 10-inch-square glass plate, dusted with a spiralling array of stellar flecks, is literally a

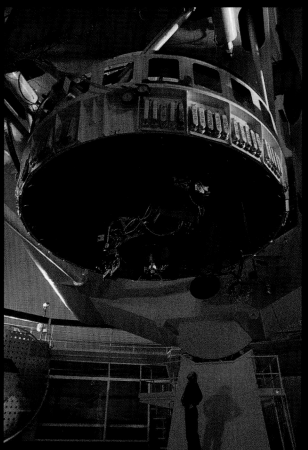

window on the universe. "The air is so clear and steady above Mauna Kea that we get sharper images of individual stars than have ever been obtained before," explains Madore. Indeed, the photograph is speckled with innumerable points. In denser regions, they merge into knots and clumps, as if piled on top of one another. The effect is an illusion: stars in the spiral arms of galaxies are several light-years apart.

Light-years define time as well as distances in the cosmos. Astronomers call it look-back time. We see everything as it was—minutes ago in the case of planets, years ago for nearby stars, millennia for more distant stars in our galaxy and billions of years ago for the most remote galaxies and quasars. Exploring the universe is a venture in both space and time.

CHAPTER EIGHT

IN SEARCH OF EXTRATERRESTRIALS

The newspaper cartoon strip *Frank and Ernest* sometimes features a pair of curious aliens scouting Earth in a flying saucer. One such excursion involved a decision about where to land. Finally, Alien A said to Alien B (they both look like humanoid dogs with antennas), "Land either in New York or in southern California so we won't look too conspicuous."

Apart from its regional earthbound jibe, the comic strip repeats once more the implicit assumption made by countless cartoonists, science fiction writers and movie producers: that aliens are indeed in the Earth's vicinity considering how to make contact with us dumb Earthlings. Judging by the box-office receipts of films such as *2010* and *Close Encounters of the Third Kind*, the idea has enormous popular appeal. But scientific opinion has been generally negative, with few outright advocates of the they-are-nearby-and-watching-us school.

I believe the objections to the concept are based more on emotional than scientific grounds because recent findings in such diverse fields as astrophysics, biology, computer technology, planetary geology and spacecraft-propulsion technology all support it in various ways. I also feel that one of the least plausible scenarios is that of creatures' signalling us with radio telescopes from a planet of a nearby star. It seems far more likely that such random contact is the last thing alien intelligences would want to achieve. But before pursuing that point, I should preface my statements with some of the theories that form the foundation of current discussions of extraterrestrial life.

The most compelling and obvious fact which supports the idea that we are not alone is the overwhelming vastness of the cosmos. Our own galaxy, the Milky Way, contains more stars than there are human beings now living on Earth plus all their ancestors. And there are billions of similar galaxies in the known universe. That is more than a billion trillion stars. Modern theories of star formation predict that at least some of the stars should have planets. What fraction of those planets would provide the necessary conditions for the emergence of life is uncertain. Nor do we know what percentage of life-bearing planets could harbour intelligent life.

Yet as small as the percentages become, we are still dealing with the residual from a billion trillion initial possibilities. True, nature is known to be profligate in these matters. Some species of plants disgorge billions of seeds in the likelihood that only one will grow to maturity. Could our universe be a garden with a billion trillion seeds and only one mature plant? I side with those who suggest that intelligence, having evolved once, should have occurred elsewhere. Either that or the life processes we see on Earth represent a remarkable event—some would say a miracle.

Setting aside the miraculous hypothesis for the moment, suppose that during the history of the Milky Way Galaxy, at least a few other intelligent technological civilizations arose on places much like Earth. What would these civilizations be like? Because the majority of stars in the Milky Way Galaxy are several billion years older than our sun, most alien life, on average, should be further evolved than we are. (Earth and sun formed virtually simultaneously 4.6 billion years ago; the oldest

Life may exist in yonder dark, but it will not wear the shape of man.

LOREN EISELEY
1957

In a distant sector of the Milky Way Galaxy, a planet about the same size as Earth but only one-tenth its age has the potential ingredients to become life-supporting. If water, the crucial ingredient, remains liquid over most of the planet for the next few billion years, intelligent creatures may arise and begin their explorations of the galaxy.

stars may have existed for as long as 16 billion years.) We are the new kids in an established neighbourhood. Unless all higher forms of life are ultimately self-destructive, there has been plenty of time for alien civilizations to develop far beyond our level of sapience.

Self-destruction is a disturbingly plausible end point for upwardly mobile intelligences. Just as a catastrophic supernova is nature's way of terminating a massive star, self-annihilation may be a natural end to life. However, I prefer to think that evolution is not inherently a dead-end affair and that some civilizations survive the bottleneck.

Interstellar Travel

Humans have pondered the question of extraterrestrial life for millennia, but the tools that have advanced the study beyond mere philo-

sophical musings have been available for only 30 years. Rockets have flung robot probes on voyages beyond the solar system and hold the promise to do the same for humans in the next century. Meanwhile, the search for worlds orbiting other suns continues with at least a dozen "discoveries" of Jovian-type planets or brown dwarfs reported in scientific journals during the last 30 years. They were all false alarms; either the finds remain to be confirmed or other astronomers refuted the data.

It is still too early to say, but the string of disappointments may be at an end. During routine studies of pulsars in 1991, sensitive radio-telescope observations showed tiny variations in the reception of pulses from one of them, pulsar PSR 1257 + 12. Analyses of the oscillations suggest that the pulsar is being affected by the gravitational pull of two planets, 3.4 and 2.8 times the Earth's mass, that orbit

Hundreds of billions of stars populate the Andromeda Galaxy, the nearest major galaxy to the Milky Way. The universe contains billions of galaxies and at least a billion trillion stars. Unless modern theories of star formation are wrong, many of those stars should have planets, yet not one planet the dimensions of Jupiter or smaller has been detected orbiting any other stars. As the search continues, planets orbiting other stars should be found before the end of the century.

the pulsar once every 67 and 98 days, respectively. These planets have circular orbits approximating those of Mercury and Venus in our own solar system. Astronomers speculate that the planets formed from debris left over after the destruction of a companion star by the supernova which created the pulsar, so this case has no direct bearing on the life question. Nevertheless, it is the strongest evidence so far of planets beyond our solar system.

Other investigations have uncovered what are almost certainly solar systems hatching—rings of dust and gas surrounding a newly formed star. The key unanswered question now seems to be not *if* planets form but how many and what size.

Our ability to begin the exploration of other worlds places us at a pivotal evolutionary juncture that creatures on other worlds either passed long ago or have not yet reached. The technological advancement from steam engine to computer over

the last two centuries took but a fraction of a second in the lifetime of the universe. It is exceedingly unlikely that another civilization in our galaxy is just beginning to explore its own planetary system. The odds are heavily stacked against two civilizations in a galaxy being anywhere close to the same technological century. To societies that evolved before us, the discovery of how to traverse the gulfs between the stars would be ancient history.

Given time and the appropriate technology, there is no reason why extraterrestrials could not travel around the galaxy or beyond. Lengthy interstellar voyages could be achieved by retarding the biological clock that controls the ageing process, by making the ship large enough to accommodate a microcosm of civilization, by sending surrogates in the form of robots (as we have done with the Voyager and Venera planetary probes) or by avoiding the time factor altogether and travelling

very close to the speed of light to utilize the time-dilation effect. Fusion or matter-antimatter propulsion systems could tap virtually limitless energy sources, permitting unbounded exploration.

But after space travel becomes commonplace and problems of health and energy abundance are solved, one question would remain for any curious civilization: does the universe harbour other rational creatures? Unless advanced civilizations are spread so thinly that there are fewer than one per galaxy, they will become aware of each other sooner or later. Either individually or collectively, the civilizations that arose before us in our galaxy must have thoroughly surveyed their interstellar environment, if for no other reason than to attempt to answer that question. Long ago, the process would have led to the discovery of life on Earth. Despite the fact that our planet is a relative newcomer on the galactic time scale, the oceans have nurtured life forms for three billion years, about one-fifth of the age of the galaxy—ample time for Earth to have been noticed as a life-bearing world.

This may sound like the background for a science fiction movie, but consider the alternatives: if aliens are not aware of us, either they must have all self-destructed or they are not interested or we have always been alone or they are there but have not yet discovered us.

The weakness of the first alternative is that it is too restrictive. To suggest that no civilizations would survive is to assume that self-destruction is a fate sealed in the cells of primordial organisms on every life-bearing world—a closed loop, like the birth and death cycles of massive stars. Similarly, the idea that all alien intelligences are uninterested in seeking other life is too narrow an assumption. It ignores the only example of technologically advanced life we know about: us.

The possibility that we have always been alone is also unlikely, given the billion trillion stars in the known universe. The origin of life would have to have been a once-only affair, which would put us in a unique position in the universe, at the apex of the cosmic schema. There is some precedent to suggest that this may be an unwise viewpoint. All previous theories that placed humans in a central or preferred position in the celestial hierarchy have proved false. There was the pre-Copernican Earth-centred universe, the post-Copernican sun-centred universe and, earlier this century, inflated estimates of our galaxy's size and special location.

A variation on the we-are-alone theme is that a steady supply of potential advanced life is always emerging in the universe but never survives due to natural but externally induced catastrophes such as severe meteorite bombardment, nearby exploding stars or passage through a dense nebula. There have been several devastating waves of extinctions on Earth that obliterated many life forms and markedly changed the course of evolution. Maybe we were just lucky—a jackpot on the cosmic roulette wheel. An intriguing idea, but it still (indirectly) implies cosmic uniqueness.

The last alternative assumes that no other civili-

*Our nearést stellar neighbour, Alpha Centauri, has not yet been ruled out as a system capable of harbouring some form of life. In this rendering, a planet experiences a twin sunset as the yellow and red suns—Alpha Centauri A and B—have just dipped below the horizon, providing a double twilight. Conditions suitable for life as we know it (carbon-based) may exist in a broad spectrum of environments, but higher forms of life might be limited to worlds with accessible land, liquid water and a stable sun. The two illustrations, **facing page**, suggest the gravitational extremes. Low mass (like that of Mars) may inhibit retention of a suitable atmosphere; high mass (over two Earth masses) might mean no continents could peek above the oceans and barges of biomass would be the only "land." On Earth, ocean creatures have existed far longer than land life, yet technological development is the preserve of land-based life. If such limitations are universal, technologically advanced life can exist only on Earthlike worlds.*

zations explore beyond the local neighbourhood. Such reasoning invokes the existence of a natural celestial quarantine that severely limits interstellar exploration by advanced civilizations. Interstellar travel does not defy any known physical laws. However, most arguments against such travel attempt to demonstrate that it would be technologically difficult and expensive and, consequently, that it would not be commonly done. But these arguments express "expense" and "difficulty" in terms of 20th-century or, at best, 21st-century technologies. Even the most perceptive visionaries, such as H.G. Wells, Jules Verne and Leonardo da Vinci, projected technological progress only a few centuries. No one alive today can possibly guess what devices will propel us (or our consciousness) to the stars. As noted science fiction author Arthur C. Clarke once said: "Any sufficiently advanced technology will seem to us like magic."

The Quest for Signals

Interstellar travel is the most direct way to seek life on other worlds. Nothing can equal actually taking a close look. And firsthand investigation is probably the only way creatures less technologically developed than humans could be detected. Searching for alien signals with radio telescopes will yield results only if other civilizations are staying at home making long-distance calls to their galactic relatives via radio frequencies. Radio-search advocates not only admit this but have enshrined the concept in what Princeton University physicist Freeman Dyson calls "the philosophical discourse dogma." He says they assume "as an article of faith" that higher civilizations communicate by radio in preference to all other options.

Prior to 1959, there was no dogma on the search for intelligent extraterrestrials. All references to the subject were found between the covers of science fiction books and magazines where interstellar spaceships were an integral part of the picture. Then, almost overnight, everything changed. It became scientifically respectable (or, in the view of some, tolerable) to talk about contacting aliens because the prestigious journal *Nature* published an article by physicists Philip Morrison and Giuseppe Cocconi describing the relative ease of transmitting radio signals across the galaxy and how it might be possible, using radio telescopes, to detect such signals from other intelligences. Since then, hundreds of articles and dozens of books embellishing the concept have been written.

The first actual search, by astronomer Frank Drake, was conducted a few months after the Morrison-Cocconi article appeared. Drake, along with Cornell University's Carl Sagan, went on to become one of the subject's chief spokesmen. Recognizing that interstellar spaceships and radio searches are concepts in conflict, Drake advanced the argument of the extreme difficulties and the enormous expense involved in travel between the stars. But that was in the early 1960s. Since then, dozens of serious research reports and proposals

have been published detailing various interstellar propulsion systems that are considered reasonable extrapolations of late-20th-century technology. That weakened the rationale for the radio searches, and by the mid-1970s, a significant number of astronomers were calling for a reassessment of the assumptions. In the ensuing discussions in scientific journals, those interested in the question have split into two camps: the agnostics and those who stand by the dogma. Scientists from one group are seldom invited to SETI (search for extraterrestrial intelligence) conferences organized by the other.

Alien Visions

Radio-search advocates have one powerful comeback: if we do not listen, we will never know for sure. I do not object to someone's monitoring the cosmic radio dial; however, the public's perception of the activity is that researchers are attempting to eavesdrop on alien conversations. Yet that is not what the radio quest is all about. Even if the galaxy were humming with alien communications by radio, we almost certainly would not intercept any of them unless the transmissions were outrageously extravagant in signal power, which goes against the initial argument about the efficiency of radio communication. Signals directed from point A to point B in the galaxy would be undetectable unless we happened to be precisely between the two points—an enormous improbability. Our radio telescopes are far too weak to eavesdrop on conversations not focused in our direction.

Could we recognize other kinds of signals? Earthlings have broadcasted their existence for about 60 years through radio, television, radar and other electronic transmissions that escape into space. These all flood out from our planet at the speed of light and now form an expanding bubble 60 light-years in radius. But could we detect the noise from another civilization? According to a 1978 study conducted by Woodruff T. Sullivan and his colleagues at the University of Washington, existing equipment would be able to pick up electronic "leakage" of television- and radar-type emissions from only a few dozen light-years away—a radius that includes less than 200 stars. In the decade since that study, the sensitivity of our receivers has advanced to extend the detection radius to include thousands of stars, although leakage from a planet like ours would be swamped within a few hun-

dred light-years by the static of natural cosmic radio sources. In any case, such leakage would not last for long, according to Carl Sagan. "I think there is just a 100-year spike in radio emissions before a planet becomes radio-quiet again," he said in a published interview a few years ago. He pointed out that communications on Earth are rapidly moving from brute-force broadcasting to narrow-beam transmission, satellite dishes, fibre optics and cable television—all of which are relatively low power. Sagan's assessment: "The chance of success in eavesdropping is negligible."

Despite the frequent use of the word "eavesdropping" in connection with the SETI programmes, most of the searches have been limited almost entirely to finding either a superpowerful omnidirectional beacon or a narrow-beam signal intentionally directed toward us. Such a search strategy therefore assumes that extraterrestrials either know about us or are pumping a colossal signal in all directions in the hope that somebody is listening.

A number of thorny questions emerge from all of this. How long would a civilization continue to broadcast an omnidirectional signal? Since it would take hundreds or thousands of years for a response, would everybody be listening and no one sending? And why would an alien civilization want to broad-

cast, not knowing who would intercept the message, what culture shock it would cause or who might be around on their home planet to receive a reply centuries or millennia in the future? If a signal is intentionally beamed in our direction, the senders must know we are here. Is the universe so perverse in its structure that it constricts intelligences to a destiny of exploration by megaphone? Is it possible that extraterrestrials might consider radio transmissions to be as quaint as we regard smoke signals or jungle drums? Throughout human history, many methods of communication have been relegated to the technological scrap heap.

Nevertheless, approximately 50 radio-telescope searches have been undertaken during the last 25 years. Since 1984, a refurbished radio telescope in Massachusetts has been scanning for artificial interstellar signals 24 hours a day, a project privately funded by The Planetary Society, a California-based organization of enthusiasts interested in continued exploration of the planets and beyond. Meanwhile, a U.S. government-sponsored search has begun using Space Agency radio telescopes during times when they are not tracking spacecraft. It is expected to be the most powerful celestial quest ever attempted.

University of Maryland astronomer Ben Zucker-

Two marginal environments for life: a planet of a close double star, **above left**, *and the atmosphere of a gas-giant planet,* **above right**. *Close double stars often affect each other's evolution through matter transfer, as in the system RW Persei, illustrated here. To avoid this, stars like the sun would have to be at least 10 diameters apart, although in such cases, a planet at the Earth's distance could not have a stable orbit over long time periods.* **Facing page:** *a radio telescope scans the distant stars.*

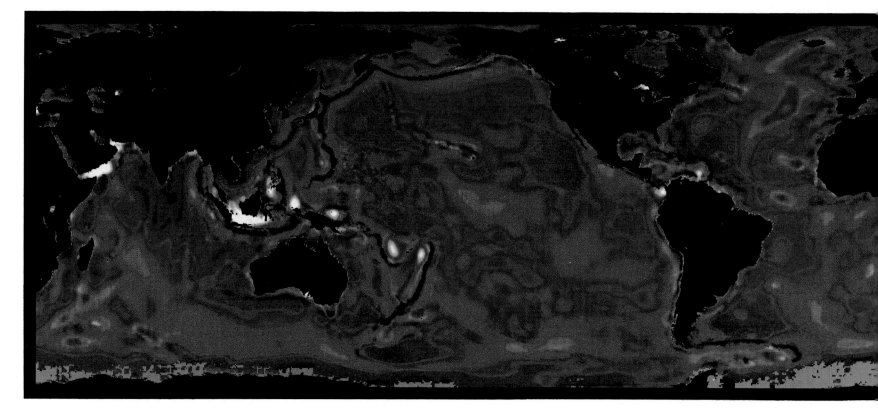

man, who conducted the largest survey prior to The Planetary Society's effort, now thinks the chances of such searches succeeding are so remote that he is no longer pursuing the activity. The late Iosif Shklovskii, an eloquent search enthusiast who coauthored *Intelligent Life in the Universe* with Carl Sagan in 1966, also grew disenchanted a decade after the book was published.

Countering the pessimism is Sagan himself, a longtime advocate of the quest for an intentional signal from another civilization. "Since the transmission is likely to be from a civilization far in advance of our own," he wrote in 1978 in an article published in *Reader's Digest*, "stunning insights are possible, [including] prescriptions for the avoidance of technological disaster. Perhaps it will describe which pathways of cultural evolution are likely to lead to the stability and longevity of intelligent species. Or perhaps there are straightforward solutions, still undiscovered on Earth, to problems of food shortages, population growth, energy supplies, dwindling resources, pollution and war."

Frankly, I find such pronouncements rather surprising. Suggestions that messages from aliens will solve all our problems not only are presumptuous but also smack more of religion than of science.

Surely superior intelligences in the universe have graduated to something more creative than operating broadcasting stations to send galactic versions of the Ten Commandments to the heathens.

Our own experience on this planet suggests that contact between technologically imbalanced cultures is bad news for the number-two culture. Painful assimilation and cultural decimation seem to be the products of such contact, even with the best of intentions. Maybe there is some biogalactic law among advanced civilizations that forbids direct intervention with primitives like us. In any event, technologically superior intelligences could probably make themselves completely undetectable, leisurely learning all they care to know about life on Earth without our being aware of the scrutiny.

Their physical appearance may likewise be unrecognizable to us. They may once have passed through an evolutionary phase when they resembled humans, but the lizard men and the gaggle of other humanoids from the *Star Wars* and *Star Trek* movies are probably not who (or what) we will meet. Evolution beyond humanlike form is as inevitable as our ascent from our reptilian ancestors. One billion years ago, the highest form of life on Earth was the worm. An alien intelligence one bil-

virtually indestructible semi-immortal arrays of silicon or its evolutionary successor. To such a form of intelligence, time would have a totally different meaning. With no finite life span to impede time-consuming activities such as interstellar travel, millions of years could be spent in exploration. Voyages to countless star systems would present no problems for a brain living in a computer. To such travellers, emerging intelligences like ours would be fascinating biological crucibles. Occasionally, they might look in on Earth to glimpse the latest tribal squabble and wonder when we will emerge to seek our place in the galactic community.

What About UFOs?

If aliens can travel here, then where are they? Where are their artifacts, their spaceships, their supply bases? So far, there is not a shred of hard evidence that extraterrestrials exist, either in our vicinity or in deep space. But absence of evidence is not evidence of absence. A century ago, it would have been impossible to identify a modern reconnaissance satellite—able to spot a human shadow in a playing field from an altitude of 120 miles—as anything more than a mysterious moving dot of light tracking across the night sky. Even that could be concealed by coating the spacecraft in ultralow-reflective material. Because there could be millions of years of technical advancement between Earthlings and aliens, it might be a simple matter for them to remain undetected. Yet there have been thousands of reports of strange aerial phenomena—unidentified flying objects, or UFOs—that have been cited as proof that extraterrestrial devices are exploring Earth.

I delved seriously into the subject after I saw a UFO in 1973 that was witnessed by 18 other people. I was teaching a class in astronomy at a planetarium in Rochester, New York. We were outside identifying constellations when we saw a formation of lights pass silently almost overhead, then veer off and swoop toward the horizon. It is not impossible for an aircraft or a group of aircraft to behave in this manner, but I had never seen that kind of formation nor the type of lights we observed.

What made the sighting even more baffling was being told when I called the community relations officer at the air force base responsible for the airspace over Rochester that no air force vehicles were anywhere near the area at the time in ques-

lion years ahead of us on the evolutionary ladder could be as different from us as we are from worms.

On Earth, the next stage beyond humans will probably be computer, not biological, evolution. In some ways, computers can be regarded as a newly emerging form of life, one built on silicon, rather than carbon (the basis for all biological life). Silicon is chemically similar to carbon in some respects, but structurally, it is relatively invulnerable to damage. A silicon computer "brain" can have unlimited size and capacity, whereas human brains have not increased in size for at least 75,000 years.

Continuing miniaturization of computer components could lead to a synthesis between humans and nonbiological computer intelligences. For example, a minicomputer might be surgically implanted in the human brain. One would merely think a question in a manner that the computer could understand, and the answer would be provided as a conscious thought. Research is now under way for surgically implanting computers in paraplegics (although not directly brain-connected).

Theoretically, there is no limit as to how far this technology could progress. Perhaps bodies of bone and flesh are already redundant in the universe, and advanced civilizations have become

Facing page: *Greek scholars of the third century B.C. were aware of the Earth's spherical shape, but to them, a view such as this could only be regarded as magic. Here, the ocean bottom is mapped by satellite and computer-processed in false colour. Other satellites, used primarily for military reconnaissance, can detect the shadow cast by a human. But a technology 20 centuries in advance of ours could have methods of remote surveillance that would appear to us as magic.* **Left:** *A planet several times the mass of Jupiter orbiting a red dwarf star could be the most common type of planetary system in the universe. Red dwarfs are the most prevalent class of star, and because gas will be present in the primal nebula of any newborn star, giant planets should be able to form even if rocky planets like Earth cannot. (Bright flecks on the planet's night side are lightning flashes like those seen on Jupiter by the Voyager spacecraft.)*

tion. A controller at the local airport said the same thing. Only through persistent sleuthing did I finally discover that what we had seen was, in fact, a small fleet of experimental helicopters, top secret at the time, a version of which was subsequently featured in the movie *Blue Thunder*.

If I had been unable to take the time and effort to research the case properly, I would have been left with a lifelong UFO story to tell. That is not to say every UFO which has ever been seen can be explained in a conventional manner, as some debunkers would have us believe. A small number of cases have been thoroughly investigated and remain a mystery. These may represent a hitherto undiscovered phenomenon, perhaps a rare atmospheric electromagnetic effect. There are several exotic explanations. Extraterrestrial spaceships capture the imagination because they are the most dramatic of the possibilities.

Some accounts of UFO and flying-saucer sightings have become modern legends. They include explicit descriptions of alien creatures and, in a few cases, bizarre tales of aliens that allegedly abducted humans for hours or days. I became sufficiently intrigued by such reports that I interviewed some of the "abductees." Despite the sincerity of these people, I remain unconvinced—and I was ready to be convinced. I have no bias against the ability of extraterrestrials to traverse the gulfs between the stars to explore Earth, but there are too many inconsistencies. Why, for instance, would they allow themselves to be seen as UFOs—elusive, yet hinting at so much—playing peekaboo with the natives? Are we to conclude that aliens choosing to venture close to Earth are *almost* clever enough to go undetected, but not quite? I think not.

Over the years, I have investigated many UFO sightings and was, for a time, a consultant for a major UFO study, which forced me to give serious thought to the matter. I have concluded that the investigation of UFOs is not a fruitful avenue to pursue in the hopes of contacting our cosmic cousins, although it may prove of value in some other way. In the late 1960s and early 1970s, discussions of UFOs entered the pages of scientific journals and engaged the interest of a cross section of researchers ranging from astronomers to psychologists. But that era seems to have ended. UFOs have virtually disappeared from the pages of scientific publications and are limited almost exclusively to sensationalistic supermarket tabloids.

Contact: Alien Communication

The standard science fiction stories of aliens invading Earth represent, in my view, the least likely scenario. Cultures more advanced than ours will be benign—they will have learned to live with themselves. Aggressive creatures prone to squabbling and warfare would be weeded out by natural selection: they would destroy themselves. If by some fluke they did not, they would be kept in check by even more advanced civilizations acting as galactic police. Otherwise, we would have known about them by now, because creatures with the will and the power to enslave or destroy us would have done so long ago.

We may be a celestial nature sanctuary that has reached a critical evolutionary stage. For the first time, the spectre of self-annihilation looms as a possibility. We would be of prime interest to alien intelligences. To understand why, consider the opposite point of view. When we have the ability to travel to other solar systems, the most exquisite discovery we could make would be other living creatures. Even recognizing a lowly bacterium on another world will be a momentous event, telling us at last that we are not alone. But that revelation would pale in comparison with the detection of a world harbouring creatures which might eventually evolve to contemplate their own existence. In any conceivable circumstance, thoughts of conquest or destruction of such a crucible would be totally pointless.

The most valuable thing we Earthlings have to offer advanced aliens is ourselves in our natural state. We are a biological and scientific oasis, a living example of intelligent beings climbing the evolutionary ladder. We are at the crucial stage where we will either become spacefaring creatures or plunge ourselves into a catastrophe of our own making. Whatever the outcome, I believe our cosmic relatives are aware of us and are observing our progress with interest.

Passive observation and nonintervention are the only approaches that would pay reasonable dividends for extraterrestrials. Despite the clamour in recent years to eavesdrop on extraterrestrials with radio telescopes, the odds favour the belief that the aliens already know about us and are silent observers. We will remain unaware of them until they are ready to talk. Contact will be made at a time and in a manner of their choosing, not ours.

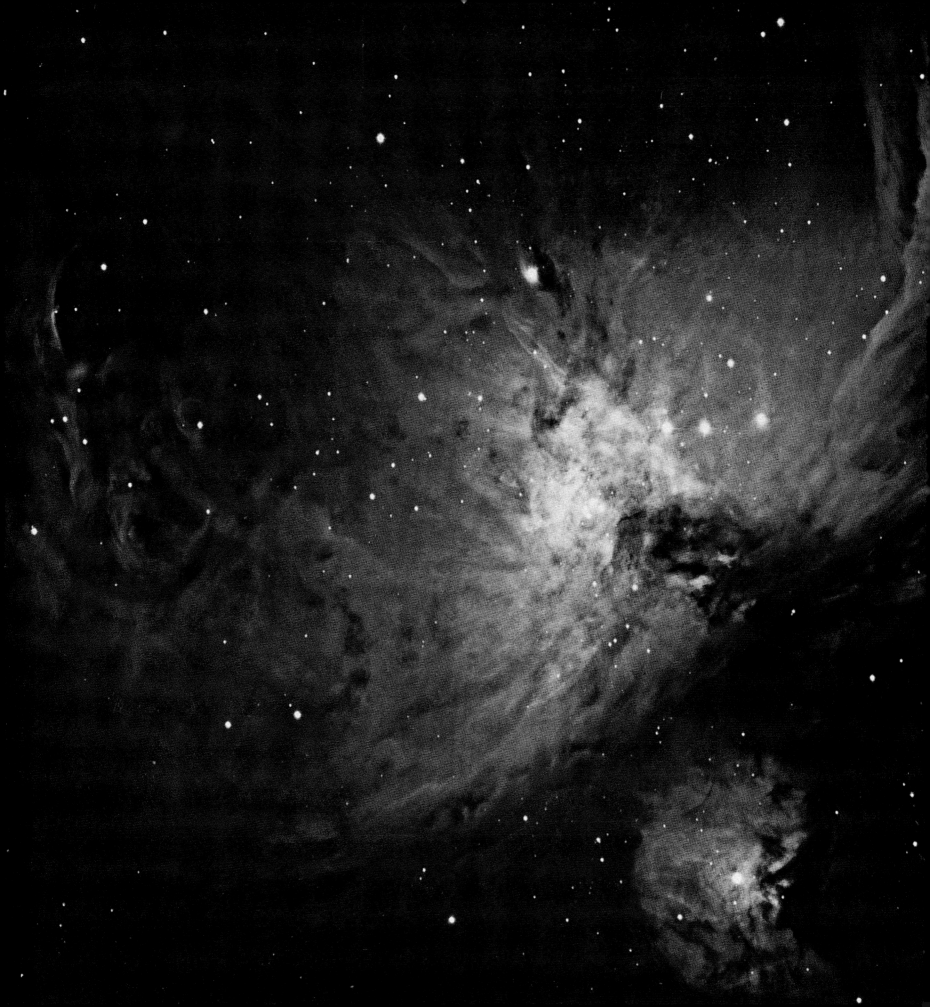

Questions of origins and destinies have always held deep fascination for humans, as evidenced by the principal preoccupation of virtually every prophet and religious doctrine. Cosmology deals with the same concerns on a cosmic rather than a human scale, although theological overtones are not entirely absent either.

The creation of our universe at a specific instant (actually, time as we know it began at that moment) seems well established now. But the universe's future has yet to be determined. There are three possibilities: the expansion now under way will continue forever (an open universe); the expansion will eventually be halted by the universe's overall mass, and a collapse will ensue, culminating in the Big Crunch (closed universe); or the mass of the cosmos is precisely tuned—as some inflationary theories suggest—so that the outrush slows to a final equilibrium, neither expanding nor contracting (flat universe). Deciding among these fates will have to wait until the invisible-mass enigma is solved, a problem that should keep astronomers busy into the next century.

In the meantime, some astronomers have fleshed out the details of the three scenarios of the universe's ultimate end. A closed universe is clear-cut. About 30 billion years after the sun becomes a white dwarf, the expansion will cease. Like a movie running in reverse, the galaxies will begin to approach one another. After some 50 billion years, the infall will reach a climax, with the galaxies rushing toward each other and merging into a fierce fireball of supernovas and superquasars. Nothing can prevent total collapse in such a situation, and the entire affair will swallow itself in a black hole the mass of the universe. All that we know today will be gone. Time and space will no longer exist.

Some cosmologists have sidestepped the finality of the closed universe by suggesting the collapse of our universe will be an event that will provide energy for the creation of another. Our universe then becomes one bead in an infinitely long string of birth, death and rebirth. The oscillating-universe hypothesis has great appeal because it embodies a dynamic, evolving universe with a definite life span. Yet it never dies; instead, it reincarnates, phoenixlike, from its own ashes. But in recent years, certain theoretical difficulties have clouded the oscillating-universe idea. Despite efforts by numerous cosmologists, not one has offered a convincing reason why the collapse will not be the final curtain for our universe.

Less tidy than the life cycle of the closed universe is the final blackness that awaits both the flat and open scenarios. Neither has a definable end; the universe just fades away. Galaxies will continue to manufacture stars from the dust and gas within them for several hundred billion years. But by 10 trillion years from now, only the least massive red dwarfs will remain. If Earth escapes being vaporized during the sun's red-giant phase, the sky of the distant era will be as black as the inside of a cave. The corpse of the sun will have cooled from a white dwarf to a dark, dense lump—a black dwarf —undetectable except for its gravity, which will continue to hold Earth in its orbit.

One hundred trillion years in the future, all the red dwarfs will cease to radiate visible light, and the universe will be a void, even telescopically. Matter will be almost entirely locked in stellar corpses: black holes, neutron stars and black dwarfs. The remainder will be in the form of planets, asteroids and comets. If current predictions in subatomic physics are correct, most of the universe's protons will have decayed a billion trillion trillion years from now, causing everything except black holes to decompose. Only the black holes will remain, leaving the universe as a desolate void punctuated with invisible gravity whirlpools.

On the time scale of the universe as we know it— where ages are measured in billions rather than trillions of years—black holes are immutable. But given enormous spans of time, they, too, will decay through a process called quantum tunnelling. A black hole the mass of the sun would take a million billion trillion trillion trillion trillion years to snuff itself out. More massive black holes would take longer, but eventually, the universe would become an exceedingly tenuous mist of subatomic particles and radiation far closer to a vacuum than intergalactic space is today. In this bleak future, the temperature everywhere would be less than one-trillionth of a degree above absolute zero.

Cosmology can say nothing about the fate of life over the universe's history. Isaac Asimov's provocative 1956 science fiction short story "The Last Question" (which he describes as his finest work) offers an intriguing prospect. In the remote future, long after the last star has winked out, all intelligences in the universe have combined into one omni-intelligence that has no physical form but

EPILOGUE: HOW THE UNIVERSE WILL END

Like a colossal flower hanging in the fabric of space, the Orion Nebula is a blossom of light on a vast cloud of gas and dust far larger than the area pictured, which is more than 25 light-years across. The nebula's illumination stems from four massive stars in a compact group called the Trapezium, seen just right of centre. A family of new stars, the Trapezium among them, is being born here, renewing the cycle of stellar birth and death that has been under way throughout the universe since the Big Bang 15 billion years ago.

133

pervades the very interstices of space. Seeing the final black death of the universe everywhere, the intelligence decides there is no alternative but to rework it into a new universe.

The universe knows nothing of our notions of a creator or of our quest for order and our distaste for a cosmos that fades into oblivion. Or does it? Some cosmologists have suggested that the universe only exists because we are here to perceive it. They say a universe with properties that do not allow stars, planets and life to emerge cannot exist. In 1985, I attended a three-day international conference where these ideas were discussed by astronomers, anthropologists, philosophers and theologians. I came away mumbling that inevitable line from low-budget science fiction movies: "There are some things man is not meant to know."

A few days later, on a tranquil summer evening, I spent several hours in the backyard with my favourite telescope. The sky was wonderfully black. Rivers of stars in the Milky Way paraded across the telescope's field of view. Saturn's rings were sharp and clear and as compelling as ever. In the Virgo cluster, nine galaxies appeared in one field, six in another. I found myself returning to the questions debated during the conference. There is no escaping it. We *need* to know how we fit into the big picture. Now that we have a reasonably accurate idea of the extent and content of the universe, our significance in it becomes the ultimate quest that will drive our descendants among those distant stars.

Perhaps our current cosmological concepts will seem as quaint to astronomers of the 22nd century as those of the 17th century seem to us. Two hundred years ago, the universe of galaxies was totally unknown. Our own galaxy's shape and dimensions could only be guessed at. Two hundred years from now, all aspects of the universe's evolution from birth to death will be mapped in detail. Though much of what we know today will become the foundation blocks for the more far-reaching cosmologies of the future, some of today's theories are destined to become historical footnotes. Nobody wants to devote a career to a long-term loser. But

it has happened many times. What is "hot" now may be irrelevant in a decade. In 1985, cosmologist P.J.E. Peebles predicted that by 1990, the inflationary-universe scenarios would be swept aside by a more comprehensive theory of the primal universe's evolution. For every researcher who agrees with Peebles, though, another can be found who maintains that the inflationary concepts are a breakthrough comparable to the discovery of the cosmic background radiation in 1965.

One of the alluring aspects of modern astronomy is that several significant revolutions in thought as well as many new discoveries can occur in a single lifetime. But that is a recent development. From the dawn of civilization until about 20 generations ago, Earth was regarded as the focus of creation, the sun, moon and stars as magical phantom lights on the crystalline dome of heaven, somewhere out of reach. To our forebears, the universe was nothing more than a giant planetarium, its machinery mysterious, its meaning shrouded in myth and superstition. For centuries, nothing happened to alter the situation. Upstarts who challenged the status quo were quickly silenced.

Today, the opposite situation prevails. For the first time in human history, it has become impossible for a citizen of planet Earth to digest the whole scope of astronomical knowledge. Trying to keep up with current discoveries is like swimming upstream: with great effort, one can do it for a short time, but inevitably, it overwhelms the best attempts. And the questions flow just as rapidly as the answers. Maybe it will always be so. If not, it means we live in a very special time. One thing that can be said for certain is that the past few decades are the most remarkable in the history of astronomy. Through space-probe investigations, we have come to know the planets in a way unimagined a generation ago. Cosmology has graduated from arm-waving speculation in the 1920s to a fairly rigorous science with a firm foundation of facts upon which theories can be built. We live in a golden age of astronomical discovery—perhaps *the* golden age. And the best part of it all is that we are here to experience it firsthand.

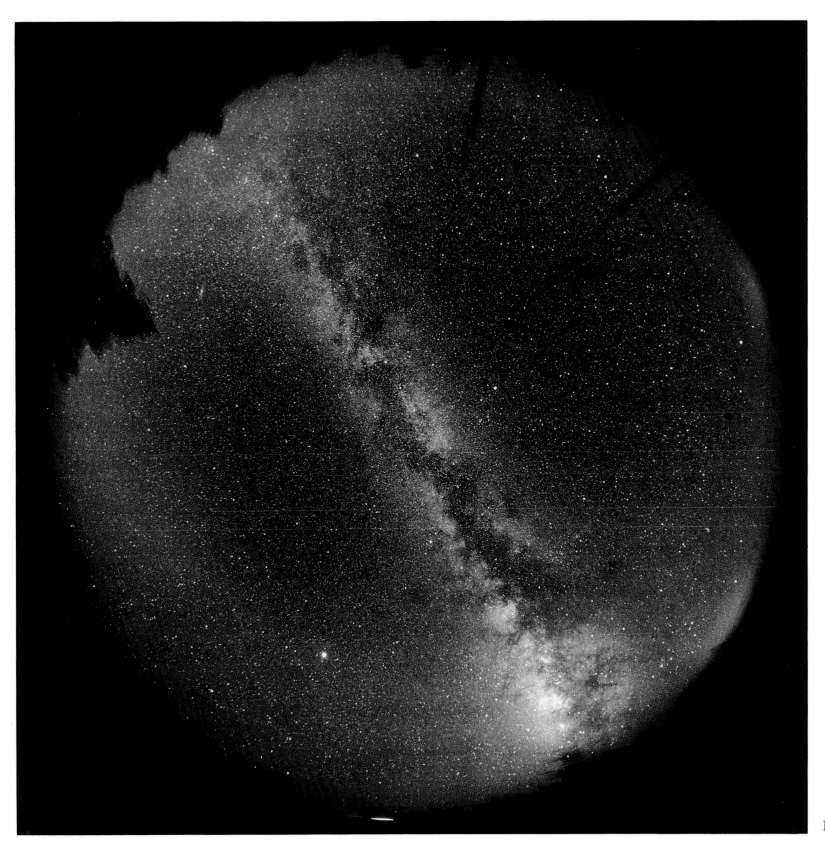

No aspect of astronomy catches people's interest as much as something new about the Big Bang. Humans' natural curiosity about origins in general seems to be ignited by the idea that we might someday know exactly how the universe began. In the late 1980s and early 1990s, new discoveries affecting the Big Bang reached the evening newscasts. This update focuses on the most important recent developments in Big Bang cosmology. It complements, rather than supersedes, the cosmology discussions in Chapter 7.

The Big Bang concept is supported by three central pillars. The first is the expansion observed as the galaxy redshifts. The second is the cosmic background radiation, the remnant radiation still swimming around space that dates back to 300,000 years after the Big Bang, when the universe had expanded and cooled enough to allow energy to flow through space without running into free electrons. The third is the observed proportion of hydrogen, helium and other elements in the universe. The intense heat of the Big Bang is the only way the lighter elements could have been created in the proportions observed, at least according to our present understanding of nuclear physics.

The problem in recent years centres around explaining how 100 billion galaxies emerged from the soup that emitted the background radiation and, further, how those galaxies were assembled in the clusters that still exist today. These enormous aggregations of galaxies, some of them more than a billion light-years across, extend throughout the cosmos in a vast three-dimensional netting that is just beginning to be mapped. For the past decade, the leading theory explaining the formation of galaxies and their evolution into immense clusters has been the cold dark matter (CDM) theory. It is this theory that is most under scrutiny, rather than the concept that time, space, energy and matter emerged simultaneously from an embryonic point in the Big Bang (although that, too, is disputed by a small band of cosmologists, as I will discuss later).

Cold Dark Matter

The CDM theory suggests that all the potentially visible stars, nebulas and galaxies in the universe combined make up only 10 percent of what's out there. The rest is invisible theoretical matter (cold dark matter) created in the Big Bang. The CDM theory rests on the fact that there is more to the universe than meets the eye. The motion of galaxies with respect to one another shows that they are being influenced by the gravity of invisible matter. The CDM theory was developed to explain the invisible matter.

In principle, the CDM theory makes eminent sense, but most versions of it have trouble explaining how the galaxies became arrayed into long chains. This conundrum came glaringly into focus in 1991, when a team of researchers announced the results of the most comprehensive galaxy survey ever conducted. The survey revealed that hundreds of small clusters of galaxies, like the one in which the Milky Way Galaxy resides, are themselves components of superclusters which are longer than anyone expected—in some cases, billions of light-years long. They are arrayed through deep space like island archipelagos. On the largest scale, the universe is a loose webbing of these galaxy ribbons, knotted here and there where the galaxies clump more densely. This is an arrangement the standard CDM theory does not predict.

Another CDM puzzle is the mysterious centrepiece of the theory: cosmions, the cold dark matter. Cosmions are theoretical subatomic particles produced when the universe was very young, dense and hot—just a mere fraction of a second after the Big Bang. Cosmions do not produce or absorb light or any other type of radiation. There could be enormous numbers of them. They could be the dominant matter in the universe by a 10-to-1 ratio, but they would remain completely invisible except for their overall gravitational influence.

Even though nobody has detected a cosmion, that has not stopped theoreticians from predicting what a universe full of them would be like. The answer is that the universe would be like the one we live in—almost. In order for the cosmions to clump together enough to collect sufficient mass to seed the formation of galaxies, the background-radiation temperature should be lumpy, indicating that the cosmions were gathering right from the start. As of early 1992, no such temperature lumpiness had been detected in the background radiation.

COBE to the Rescue

In April 1992, like the cavalry charge to save the day in the old Westerns, analysis of readings from NASA's Cosmic Background Explorer (COBE) satellite supplied welcome news to beleaguered cosmologists. Measurements of tiny variations in the background-radiation temperature were hailed as a crucial missing link supporting theories that attempt to explain the evolution of the universe from an atom-sized speck an instant after the Big Bang to the cosmic ocean of galaxies we see today.

COBE was explicitly designed to measure variations in the temperature of the background radiation to an accuracy of 1/100,000 of one Celsius degree. At that level of detection, COBE showed, for the first time, that the primordial universe was not perfectly smooth, as it had previously appeared to be. The temperature differences are actually incredibly tiny—just 3/100,000 of a degree variation from 2.735 degrees above absolute zero—but enough to indicate that the primordial universe was already showing signs of clumping.

The primordial patches are not the seeds of the galaxy clusters—they are too large for that—but the next generation of experiments, some of which can be conducted from the Earth's surface at the South Pole, will reveal 10 times the detail. Cosmologists will then be able to pin down which of several competing CDM models of the universe's evolution fit both the COBE results and the observations of the new galaxy-cluster chain.

The Plasma Universe

The COBE findings have not deterred a small but growing number of physicists who have jumped off the Big Bang wagon and suggest that the Big Bang theory may not be the best explanation for the origin and evolution of the universe. Although still a distinct minority, the doubters are rallying around a new concept, called the plasma universe, which they say explains some nagging questions more neatly than the Big Bang.

For instance, they point out that the vast chains of galaxies must have taken at least three times longer to develop than the entire age of the universe as predicted by the Big Bang. Another flaw, say the plasma scientists, is the dark matter—the cornerstone of the CDM theory—which has never been seen. One further problem is that the oldest stars appear to be 14 to 16 billion years old, yet the latest estimates of the age of the universe in the Big Bang scenario are in the 12-to-15-billion-year range.

All these discrepancies, say the Big Bang skeptics, can be explained in the plasma-universe concept, which proposes that the universe always ex-

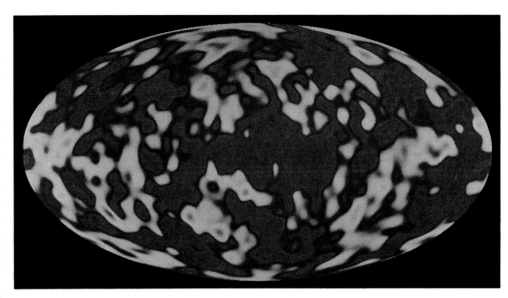

isted and its structure is dominated by electricity and magnetism, rather than gravity. A state of matter called plasma, in which atoms are separated into their component electrons and ions, is rare on Earth but occurs naturally in space. Plasma is alive with electric and magnetic fields. On very large scales, according to the plasma-universe proponents, the universe's electric and magnetic fields overpower gravity to organize matter into galaxies and stars. This is not as farfetched as it might seem, because the sun, for example, is entirely plasma. Something as close to us as the aurora is also a plasma phenomenon.

One strength of the plasma universe is that it predicts the vast billion-light-year-long ribbons of galaxies that apparently pervade the cosmos. Plasma that naturally occurs in space would, by its nature, form vast filaments that could array the galaxies like strings of Christmas lights. Another pillar of the Big Bang, the cosmic background radiation, can be explained just as easily as the scattered energy of previous generations of massive stars, say the critics. The plasma-universe advocates even have an alternative explanation for the redshift, the foundation of the expanding-universe concept, which in turn is the backbone of the Big Bang theory.

The plasma universe is not a new idea. Hannes Alfvén, a Nobel prizewinning physicist, has been promoting the hypothesis for decades, although he has had few converts until recently. But the Big Bang is still supported by the vast majority of astronomers, and the COBE results have put them more at ease than they were a year or two earlier.

Hailed as one of the most important scientific finds of the century, this false-colour image shows slight temperature variations in the cosmic background radiation as measured by COBE, the Cosmic Background Explorer satellite. The variations are tiny, only 3/100,000 of one Celsius degree, but they indicate that just 300,000 years after the Big Bang, matter was not spread evenly over the universe. The slightly denser regions, shown light blue, appear a bit cooler than the less dense regions, pink. Astronomers think those tiny temperature ripples are the primordial seeds for vast galaxy clusters that would arise one or two billion years later. If the temperature ripples had not been found, the development of galaxies from the Big Bang would be tough to explain.

Sometime during the spring of 1955, a time when I still had more than a year to wait to become a teenager, my father took me to the big city library. This was done partly to keep me quiet. I had read all the astronomy books in the school library and wanted more, but there were no other nearby sources.

After we passed through the library's heavy oak-and-glass doors, Dad pushed me in front of the librarian seated at the information desk. I looked at the "Quiet Please" sign, hesitated, then whispered, "Do you have any books about planets and stars?" The librarian took off her glasses, stood up, glided to the correct aisle and pointed to a shelf.

"A whole shelf of astronomy books!" I thought. This was too good to be true. And, after a closer look, I saw that indeed, it was too good to be true. The selection of introductory astronomy books —those with titles like *A Beginner's Star-Book*— totalled no more than a dozen well-worn volumes. I began examining them as my father went over to the magazine racks to look for *Motor Trend*.

One of the books had no dustcover, just a naked, navy-blue hard cover with the title *The Conquest of Space* down the spine. But when I opened it, I experienced one of those great moments of discovery —the ones that become etched in the brain for life. The full-colour illustrations of alien planetary landscapes had a depth and realism unlike anything I had seen before. The illustrator was Chesley Bonestell. His paintings showed Mars as seen from its moon Phobos, Jupiter from Amalthea, Saturn from Titan and dozens more that propelled me into a new universe.

It took a year of saving and pestering before I had my own copy of *The Conquest of Space*. And I was not alone in wanting a personal copy. First published in 1949, it became the best-selling astronomy book to that time, and it is now a prized collectors' item. *The Conquest of Space* was a unique collaboration of two visionaries: Willy Ley, a German rocket-scientist-turned-science-writer, and Chesley Bonestell, an architectural illustrator who was 56 when he painted his first piece of space art.

Ley and Bonestell produced several similar books in the 1950s, but *The Conquest of Space* was significant because it combined, for the first time, realistic and technically accurate colour illustrations of planetary exploration. An entire generation of young astronomy and space-exploration buffs (like me) was profoundly influenced by this book.

Bonestell gained a much larger following when his paintings were used to illustrate a 1952 series of articles in *Collier's*, an American mass-circulation magazine. Written by Wernher von Braun, Willy Ley and others, the series described the steps required for human expeditions to the moon and Mars. Bonestell's visions of space stations and nuclear rockets to Mars were powerful stuff, firing the imaginations of thousands of nascent rocket scientists and astronomers. Recognizing a good idea when he saw it, Walt Disney used the *Collier's* series as the basis for several animated shorts on space travel and life on other worlds, which were eventually seen by millions worldwide.

Bonestell was the father of modern space art, and for two decades after *The Conquest of Space* was published, he remained its most visible practitioner. A few young space artists, inspired by the master but toiling in Bonestell's shadow for years, finally emerged as masters in their own right during the 1970s and 1980s, producing stunning work portraying the wealth of new scientific discoveries during the heyday of planetary exploration. Several of them—Don Davis, David Egge, James Hervat, Michael Carroll, Brian Sullivan and Victor Costanzo —contributed to *The Universe and Beyond*. The work of the book's major contributor, Adolf Schaller, set new standards of accuracy and excellence in astronomical illustration.

Most of the Schaller paintings in this book were done in the early 1980s, when the artist was in his twenties (and Bonestell was in his nineties). My favourite Schaller pieces are displayed on pages 44, 110-11 and 120. Two more of his works are reproduced here: a spectacular asteroid impact on the primal Earth and, overleaf, the cover illustration in its complete form. Schaller was a major contributor to Carl Sagan's *Cosmos* television series and was responsible for the superb renderings of Jupiter and its satellites in the film *2010*, the sequel to *2001: A Space Odyssey*.

Historical Perspective

The importance of astronomical art is, I believe, vastly underrated. Describing in words a scenario for a trip to the moon or Mars, however detailed and well crafted, pales in comparison with the same voyages illustrated in a realistic fashion. It was the images more than the words that convinced people it could happen. And after it did happen (in

the case of the moon), the photographs told the story far more vividly than did either the astronauts themselves or Walter Cronkite. For instance, the Apollo astronauts' photographs of Earthrise over the moon have not retreated into obscurity. Instead, they remain icons of the space age. It is no coincidence that the modern environmental movement began the same year (1970) that the Apollo photographs of Earth first sank into human consciousness.

This point is seldom, if ever, made by either historians or other writers chronicling the impact of space images on attitudes and perception. For example, Bonestell's various depictions of Earth-orbiting space stations (based mainly on designs by Wernher von Braun) became so implanted in the minds of a generation of engineers, designers and NASA officials that billions of dollars have been spent on the development of a space station—even though nobody knows what it will be used for. Von Braun's rationale was that the space station would be an assembly point for the construction of ships to carry crews to the moon and Mars. This, everyone now agrees, was never a practical idea.

Another example, mentioned in a different context in Chapter 2, is the legacy of Bonestell's paintings of the surface of Mars, showing a blue sky and Arizona-like deserts. They became the "Mars of the mind" and were emulated by most other space artists before the Viking landers revealed the real rock-strewn Mars with its peach-coloured sky. So ingrained was the Bonestellian Mars that the first colour image released to the news media of the surface of Mars had brown rocks and a blue sky. After checking the spacecraft's colour-calibration system overnight and realizing that the yellow sky and the orange surface were the true colours, the mission scientists somewhat sheepishly presented the "corrected" version of the photograph.

Similarly, more recent depictions of as-yet-unseen celestial vistas—pulsars, black holes, comet surfaces, alien life forms, and so on—will remain for future generations as collective mental images of what these things "should" look like. To a large degree, this is a good thing. Space art that challenges the imagination is a major reason why interest in astronomy remains so high. I suspect that *The Universe and Beyond* will still be on shelves well into the 21st century, not because of the words but because of the pictures—a salute to the space artists and their power to take us into the cosmic abyss.

Adolf Schaller's vision of a galactic starburst—the creation of billions of stars at a galaxy's heart one or two billion years after the Big Bang—seen from a hypothetical planet. As a galaxy forms, there must be a critical period of rapid infall of gas into the galactic core, which would ignite a torrent of star births. Soon after, the most massive stars would detonate as supernovas. Expelled star-stuff, shock waves from the supernovas and more infalling gas would combine in turbulent fronts that would lead to more star formation and more supernovas, the process escalating to the crescendo of cosmic fury seen here. In the foreground, a few hundred light-years from the starburst tumult, the surface of a hypothetical planet displays the exposed remains of solidified plutons (dense igneous rock) originally deep beneath the surface. The crust that once covered the plutons has been cataclysmically scoured off the planet by the ablative action of repeated supernova shock waves. The largest plutons are higher than any skyscraper on Earth. A red dwarf star, out of the scene to the left, spreads a warm glow on the landscape. Blue light from the starburst region provides a second "sun" to this barren world.

141

A BRIEF HISTORY OF PLANETARY EXPLORATION (1949-91)

1949: Publication of the book *The Conquest of Space* by Chesley Bonestell and Willy Ley containing the first realistic illustrations of planetscapes and planetary exploration. This important work impresses a whole generation of future astronomers and NASA designers.

1950: The feature film *Destination Moon* is released. Using imaginative special effects (for the time), it portrays the first scientifically plausible scenario for space exploration seen in a mainstream theatrical release.

1952: A series of beautifully illustrated articles about spaceflight by Wernher von Braun and others appears in *Collier's*, a mass-circulation American magazine. The series deeply influences thousands of readers who would later pursue careers in science and engineering.

1957: U.S.S.R. puts into orbit the first artificial satellite, Sputnik 1, on October 4. The space age is born.

1958: The first U.S. satellite, Explorer 1, is launched by von Braun's team with a Jupiter-C rocket on January 31.

1959: Luna 3 circles the moon and transmits to Earth the first photographs of the moon's far side.

1961: Yuri Gagarin, first person in space, makes a single orbit of Earth on April 12 in his Vostok 1 capsule. He lands safely in the Soviet Union.

1962: U.S. Mariner 2 probe passes Venus; first successful flyby of any planet.

1965: U.S. Mariner 4 makes first successful flyby of Mars. Sends back fuzzy pictures showing craters.

1966: Soviet Union's Luna 9 sends back the first pictures from the surface of the moon.

1967: Soviet Venera 4 spacecraft makes the first entry into the atmosphere of another planet as it descends by parachute into the dense clouds of Venus. It does not survive to the surface.

1968: Release of the landmark movie *2001: A Space Odyssey* heightens the drama surrounding the upcoming Apollo moon flights and romances a new generation of space-travel enthusiasts.

1968: Three astronauts in Apollo 8 are the first humans to orbit the moon during test of moon-flight hardware.

1969: Apollo 11 astronauts Neil Armstrong and Buzz Aldrin walk on the moon on July 20. More than one billion people watch the event live on television.

1970: Soviet Venera 7 probe lands on Venus and transmits data from the surface. This is the first successful landing on another planet.

1970: Soviet Lunokhod 1 wheeled moon rover is the first mobile robot device on another world.

1971: Soviet Mars 3 probe lands on Mars but returns only 20 seconds of data.

1971: U.S. Mariner 9 probe orbits Mars to become the first device to orbit another planet. Over the next year, Mariner 9 maps most of the planet.

1972: The Apollo 17 mission ends the U.S. moon-landing programme. In six landings (1969-72), 12 astronauts explored the lunar surface.

1973: U.S. Pioneer 10, first probe to fly by Jupiter.

1974: Mariner 10, first spacecraft to reach Mercury, photographs the planet during flyby.

1975: Venera 9 transmits to Earth the first pictures from the surface of another planet—Venus.

1976: First pictures from the surface of Mars are transmitted to Earth by Vikings 1 and 2.

1979: Pioneer 11, first spacecraft to reach Saturn.

1979: Voyagers 1 and 2 fly by Jupiter and gather the first detailed photographs of the giant planet.

1980: First detailed pictures of Saturn—from Voyager 1 flyby. Spacecraft photographs of the planet grace the covers of *Time*, *Newsweek* and many other publications.

1983: U.S. Pioneer 10 probe reaches a point farther from the sun than all the planets and, in doing so, becomes the first human artifact to leave the solar system.

1986: Voyager 2 makes first flyby of Uranus and sends back excellent-quality pictures.

1986: A flotilla of five space probes from Europe, Japan and the Soviet Union explores Halley's Comet.

1989: Voyager 2 makes first flyby of Neptune and sends back excellent-quality pictures.

1991: U.S. Magellan spacecraft maps Venus from orbit by radar.

1991: On its way to Jupiter, Galileo probe takes first close-up photographs of an asteroid.

THE SUN AND ITS PLANETS

Object	Diameter (Earth = 1)	Mass (Earth = 1)	Density (water = 1)	Distance from sun (Earth's distance = 1)	Rotation period*
SUN	109.1	332,946	1.41	—	25 to 35 days
MERCURY	0.382	0.055	5.43	0.387	176 days
VENUS	0.949	0.815	5.24	0.723	117 days
EARTH	1.000	1.000	5.52	1.000	24 hours
MARS	0.532	0.107	3.94	1.524	24h 39m
JUPITER	11.19	317.8	1.33	5.20	9h 50m
SATURN	9.41	95.2	0.77	9.54	10h 39m
URANUS	4.01	14.5	1.27	19.18	17h 14m
NEPTUNE	3.88	17.2	1.64	30.06	16h 06m
PLUTO	0.19	0.0020	1.9	29.6 to 49.3	6d 9.3h

Object	Revolution period (length of year)	Average amount of sunlight received (Earth = 1)	Surface gravity † (Earth = 1)	Known moons	Axis inclination	Minimum light-time from Earth
SUN	—	—	27.9	—	7.3°	8.3 min.
MERCURY	88.0 days	6.6	0.38	0	0.0°	4.5 min.
VENUS	224.7 days	2.2	0.91	0	177.3°	2.3 min.
EARTH	365.3 days	1.0	1.00	1	23.4°	—
MARS	687.0 days	0.44	0.38	2	25.2°	3.1 min.
JUPITER	11.86 years	0.037	2.69	16	3.1°	33 min.
SATURN	29.46 years	0.011	1.19	18	26.7°	66 min.
URANUS	84.0 years	0.0028	0.91	15	97.9°	2.5 hours
NEPTUNE	164.8 years	0.0011	1.19	8	28.8°	4.0 hours
PLUTO	247.7 years	0.001 to 0.0004	0.06	1	94°	3.9 hours

*Rotation period of the sun varies with latitude; for the planets, the figure given is period from one sunrise to the next at the equator.

†For sun and giant planets, "surface" refers to *visible* surface (cloud tops and solar photosphere).

Earth's diameter is 7,926 miles; its distance from the sun is 93 million miles.

MAJOR PLANETARY SATELLITES

Planet	Satellite	Diameter (moon = 1*)	Distance from planet surface in planetary radii	Orbit period (days)	Mass (moon = 1)	Surface gravity (as a percent of Earth's)
EARTH	Moon	1.00	58.75	27.32	1.00	16
JUPITER	Io	1.04	4.89	1.77	1.21	17
	Europa	0.90	8.33	3.55	0.65	14
	Ganymede	1.51	15.90	7.16	2.01	19
	Callisto	1.38	25.20	16.69	1.47	17
SATURN	Mimas	0.11	2.10	0.94	0.0005	2
	Enceladus	0.14	2.94	1.37	0.001	2
	Tethys	0.30	3.89	1.89	0.011	4
	Dione	0.32	5.27	2.74	0.014	4
	Rhea	0.44	7.72	4.52	0.034	6
	Titan	1.48	19.24	15.95	1.83	24
	Hyperion	0.07	23.55	21.28	?	1
	Iapetus	0.42	58.0	79.33	0.026	5
	Phoebe	0.06	213.8	550.46	?	1
URANUS	Miranda	0.14	4.10	1.41	0.001	2
	Ariel	0.34	6.52	2.50	0.01	6
	Umbriel	0.34	7.58	4.14	0.01	5
	Titania	0.46	16.17	8.71	0.12	8
	Oberon	0.45	21.98	13.46	0.11	8
NEPTUNE	Triton	0.78	22.08	5.88	0.3	18?
PLUTO	Charon	0.4?	17?	6.39	?	5?

*Moon's diameter is 2,160 miles.

MINOR PLANETARY SATELLITES

Planet	Satellite	Diameter (miles)	Distance from planet surface in planetary radii	Remarks
MARS	Phobos	13	1.77	nearest to parent planet
	Deimos	7	5.93	smallest known moon
JUPITER	Metis	25	0.792	at outer edge of Jovian ring
	Adrastea	15	0.807	
	Amalthea	105	1.52	orbit Jupiter within Io's orbit
	Thebe	60	2.11	
	Leda	10	155	
	Himalia	115	160	middle group
	Lysithia	20	163	
	Elara	45	164	
	Ananke	20	290	
	Carme	25	313	outer group
	Pasiphae	30	326	
	Sinope	20	332	
SATURN	Pan	10	1.22	in Encke gap in ring A
	Atlas	20	1.27	at outer edge of main ring
	Prometheus	60	1.30	"shepherd" for skinny ring F
	Pandora	50	1.35	"shepherd" for skinny ring F
	Janus	115	1.50	shares Epimetheus's orbit
	Epimetheus	70	1.50	shares Janus's orbit
	Telesto	15	3.89	shares Tethys' orbit 60° behind
	Calypso	15	3.89	shares Tethys' orbit 60° ahead
	Helene	20	5.27	shares Dione's orbit 60° ahead
URANUS	Cordelia	15	0.94	"shepherd" for outer ring
	Ophelia	20	1.10	"shepherd" for outer ring
	Bianca	30	1.33	
	Cressida	40	1.43	
	Desdemona	40	1.47	all 10 of Uranus's small satellites
	Juliet	60	1.53	were discovered by Voyager 2
	Portia	70	1.60	during its approach to Uranus in
	Rosalind	40	1.75	late 1985 and January 1986
	Belinda	45	1.96	
	Puck	105	2.39	
NEPTUNE	Naiad	40	0.92	
	Thalassa	50	1.00	inner six satellites discovered by
	Despina	90	1.11	Voyager 2 prior to its flyby of
	Galatia	100	1.49	Neptune in August 1989
	Larissa	120	1.96	
	Proteus	260	3.72	
	Nereid	200	58 to 405	very elliptical orbit

145

NEARBY STELLAR SYSTEMS

Name	Distance from sun in light-years	Type	Mass (sun=1)	Diameter (sun=1)	Luminosity (sun=1)	Remarks
SUN	—	main sequence G2	1	1	1	nearest star
JUPITER	—	Jovian planet	0.001	0.10	0.000000001	largest known planet
PROXIMA CENTAURI	4.24	red dwarf M5	0.11	0.11	0.00006	nearest star to solar system
ALPHA CENTAURI A	4.34	main sequence G2	1.1	1.2	1.3	A to B = 24 AU
ALPHA CENTAURI B	4.34	main sequence K1	0.9	1.0	0.36	AB to Proxima = 13,000 AU
BARNARD'S STAR	5.97	red dwarf M4	0.15	0.12	0.00044	possible companion
WOLF 359	7.7	red dwarf M6	0.10	0.1	0.00002	
BD+36°2147 A	8.2	red dwarf M2	0.35	0.3	0.0052	A to B = 0.07 AU
BD+36°2147 B	8.2	brown dwarf	0.02?	0.1?	?	existence uncertain
L726-8 A	8.4	red dwarf M6	0.44	0.1	0.00006	A to B = 11 AU
L726-8 B	8.4	red dwarf M6	0.35	0.1	0.00004	
SIRIUS A	8.6	main sequence A1	2.3	1.8	23	brightest within 25 light-years
SIRIUS B	8.6	white dwarf	1.0	0.02	0.003	A to B = 20 AU
ROSS 154	9.4	red dwarf M4	0.15	0.12	0.0005	
ROSS 248	10.4	red dwarf M5	0.13	0.1	0.0001	
EPSILON ERIDANI A	10.8	main sequence K2	0.76	0.9	0.3	
EPSILON ERIDANI B	10.8	brown dwarf?	0.003?	0.2?	0.0000001	needs confirmation
ROSS 128	10.9	red dwarf M4	0.15	0.12	0.00035	
61 CYGNI A	11.1	main sequence K4	0.6	0.7	0.08	A to B = 85 AU
61 CYGNI B	11.1	main sequence K5	0.5	0.6	0.04	
61 CYGNI A1	11.1	large Jovian?	0.01?	orbits 61 Cygni A		needs confirmation
61 CYGNI B1	11.1	large Jovian?	0.008?	orbits 61 Cygni B		needs confirmation
EPSILON INDI	11.2	main sequence K3	0.6	0.7	0.13	
BD+43°44 A	11.2	red dwarf M1	0.3	0.3	0.006	A to B = 156 AU
BD+43°44 B	11.2	red dwarf M4	0.15	0.13	0.0004	
L789-6	11.2	red dwarf M5?	0.13	0.1	0.0002	
PROCYON A	11.4	main sequence F5	1.8	1.7	7.6	A to B = 16 AU
PROCYON B	11.4	white dwarf	0.6	0.01	0.0005	
TAU CETI	11.8	main sequence G8	0.8	0.9	0.44	first object searched for ET radio signals

Distances are known within 0.1 light-years; most masses and diameters are estimates; all listed objects are invisible to unaided eye except: sun, Jupiter, Alpha Centauri (A & B appear as one star), Sirius A, Epsilon Eridani A, 61 Cygni A, Epsilon Indi, Procyon A and Tau Ceti.

One light-year equals 5.879 trillion miles.

AU (astronomical unit) = the distance from Earth to sun; 93 million miles.

THE BRIGHTEST STARS

(as seen from Earth)

Name	Apparent brightness*	Luminosity (sun = 1)	Distance in light-years	Diameter (sun = 1)	Mass (sun = 1)	Type and spectrum	Remarks
SIRIUS A	100	23	8.6	1.8	2.3	main sequence A1	white dwarf companion
CANOPUS	52	1,400	100	30	6	giant A9	
ALPHA CENTAURI A	35	1.3	4.3	1.2	1.1	main sequence G2	triple-star system (see Chapter 5)
ARCTURUS	29	115	36	23	4	giant K1	nearest giant
VEGA	26	55	26	3	3	main sequence A0	
CAPELLA A & B	25	90 & 70	45	13 & 7	3 & 2.5	giant G2 & G6	tight binary
RIGEL	24	55,000	900	50	20	supergiant B8	has tight binary companion plus distant pair
PROCYON A	20	7	11.4	2.2	1.8	main sequence F5	has white-dwarf companion
ACHERNAR	17	800	90	6	10	main sequence B3	
HADAR A	15	5,000	320	7	16	main sequence B1	
BETELGEUSE	14	8,000 to 14,000	520	500 to 800	18	supergiant M2	slowly varies in brightness and size
ALTAIR	13	10	17	1.5	2	main sequence A7	
ALDEBARAN	12	125	68	45	4	giant K5	has less massive companion star
ACRUX A & B	12	4,000 & 1,500	370	8 & 6	18 & 14	main sequence B0 & B1	binary
ANTARES	11	9,000	500	300	10	supergiant M1	main-sequence B3 companion
SPICA A & B	10	1,300 & 250	220	7 (A)	15 (A)	main sequence B1 (A)	binary
POLLUX	9	35	35	11	1.5	giant K0	
FOMALHAUT	9	14	23	2.5	2	main sequence A3	
DENEB	8	60,000	1,600	60	30	supergiant A2	most luminous bright star in the Earth's sky
BETA CRUCIS	8	5,500	470	11	17	main sequence B0	less massive companion star
REGULUS	8	160	85	4	5	main sequence B7	less massive companion star
ADHARA	7	8,000	650	25	20	giant B2	
CASTOR A1 & A2	6	12 & 12	45	3 & 3	3 & 2.5	main sequence A1 & A2	six-star system; two close pairs, one distant pair

*As seen without optical aid in the Earth's night sky; figure given is a percent, in comparison with Sirius, which is arbitrarily given a value of 100.

Distances to stars within 200 light-years are known to within a few percent; accuracy of measurement falls at greater distances.

Diameters and masses of most stars are estimates.

SOME CHARACTERISTICS OF MAIN-SEQUENCE STARS

Mass	Spectral type	Luminosity (sun=1)	Diameter (sun=1)	Central density (water=1)	Lifetime on main sequence (millions of years)
30	O8	100,000	15	3	4
15	B1	20,000	10	6	11
10	B5	5,000	5	9	20
5	B8	500	3	20	75
2	A6	17	1.7	70	800
1.5	F3	5	1.3	85	1,800
1	G2	1	1	90	10,000
0.5	K8	0.03	0.7	80	100,000
0.1	M7	0.0001	0.1	60	1,000,000

Mass	Death process	Mass of final remnant star	Probable form of remnant
30	supernova	4 to 20	black hole
15	supernova	1.4 to 10	black hole or neutron star
10	supernova	0.5 to 4	neutron star or white dwarf
5	gradual mass loss and collapse	less than 1.4	white dwarf
2	gradual mass loss and collapse	less than 1.4	white dwarf
1.5	gradual mass loss and collapse	less than 1.4	white dwarf
1	minor mass loss and collapse	about 0.9	white dwarf
0.5	gradual cooling	same	white dwarf
0.1	gradual cooling	same	brown dwarf

LIFE HISTORY OF THE SUN

Event	Age (millions of years)	Luminosity (present sun = 1)	Diameter (present sun = 1)
Contraction from nebula into protostar	0	100	50
Hot core forms from contraction	1	20	20
MAIN-SEQUENCE LIFE BEGINS			
Protostar contraction ends; nuclear fusion of hydrogen begins	70	0.6	1.0
Today	4,600	1.0	1.0
Hydrogen begins to be depleted at sun's core	7,000	1.4	1.2
Hydrogen fusion requires higher temperatures due to helium buildup	9,000	2	1.5
Hydrogen "burning" moves to a shell around helium core	10,000	4	3
MAIN-SEQUENCE LIFE ENDS			
First red-giant stage reaches maximum, and helium core ignites	10,600	1,500	50
Helium fusion in core nears maximum	10,630	100	10
Final red-giant stage begins	10,650	1,000	100
Final red-giant stage reaches maximum as helium fusion shifts to shell around core	11,000	10,000	400
Sun sheds matter as variable or planetary nebula	11,000	variable	contracting
White dwarf forms in 75,000 years	11,000	1/300	1/100
White dwarf cools to black dwarf	50,000?	0	1/100

LIFE HISTORY OF A MASSIVE STAR
(15-solar-mass example)

0 to 10,000 years	contraction of nebula, collapse of cloud and formation of star leading to ignition of thermonuclear reactions at core
10,000 to 10,000,000 years	main-sequence star: combustion of hydrogen in the core; luminosity about 30,000 times the sun's
next 200,000 years	beginning of helium combustion in core; star begins to expand from 10 to 30 solar diameters
next 700,000 years	continued expansion to 50 solar diameters; star begins to pulsate as a Cepheid variable
next 700,000 years	irregular variability sets in; expansion continues to 100 solar diameters
next 30,000 years	expands to red supergiant like Betelgeuse, 500 times sun's diameter; luminosity fairly constant throughout all above phases
sudden onset	supernova; total life span about 11 million years

SELECTED GALAXIES

Name	Distance (millions of light-years)	Diameter (light-years)	Stellar mass (billions of solar masses)	Type
MILKY WAY	—	80,000	200	spiral
LARGE MAGELLANIC CLOUD	0.17	30,000	10	barred spiral?
SMALL MAGELLANIC CLOUD	0.19	15,000	2	irregular
ANDROMEDA GALAXY	2.3	120,000	300	spiral
M32 (Andromeda companion)	2.3	6,000	3	elliptical
M110 (Andromeda companion)	2.3	10,000	8	elliptical
M33 (Triangulum Galaxy)	2.5	50,000	15	spiral
NGC 55	8	75,000	50	spiral
NGC 247	9	50,000	30	spiral
NGC 253	10	60,000	150	spiral
M81	10	60,000	150	spiral
M82	10	25,000	30	peculiar
M83	15	60,000	200	barred spiral
NGC 5128 (Centaurus A)	17	60,000	200	elliptical/peculiar
M94	19	35,000	100	spiral
M64 (Blackeye Galaxy)	20	60,000	100	spiral
M101 (Pinwheel Galaxy)	21	165,000	300	spiral
M51 (Whirlpool Galaxy)	25	60,000	200	spiral
M106	28	150,000	400	spiral
M63	30	100,000	200	spiral
M66	30	90,000	250	spiral
M104 (Sombrero Galaxy)	40	110,000	600	spiral
M74	45	130,000	400	spiral
M87	60	75,000	4,000	elliptical
M60	60	100,000	300	spiral
M77	70	100,000	800	spiral

Estimated distances to galaxies beyond M33 could be in error by several million light-years, which would affect other values in the table. All galaxies in the table are visible in binoculars or small telescopes. Stellar-mass estimates do not include "invisible" mass described in Chapter 7.

A COSMIC HIERARCHY

A selection of increasingly larger objects and structures

Name	Longest dimension	Object/Structure
ICARUS	½ mile	small asteroid
DEIMOS	7 miles	smallest planetary satellite
HALLEY'S COMET (NUCLEUS)	9 miles	large comet
CERES	600 miles	largest asteroid
PLUTO	1,500 miles	smallest planet (largest comet?)
MOON	2,160 miles	
GANYMEDE	3,268 miles	largest planetary satellite
EARTH	7,926 miles	
JUPITER	88,734 miles (11 Earth diameters)	largest planet
SUN	865,000 miles (109 Earth diameters)	nearest star
BETELGEUSE	800 solar diameters	largest nearby star
ORION NEBULA	25 light-years (25,000 solar system diameters)	nearest large illuminated nebula
HERCULES CLUSTER	75 light-years	large globular star cluster
LARGE MAGELLANIC CLOUD	20,000 light-years	nearest galaxy
ANDROMEDA GALAXY	120,000 light-years	nearest spiral galaxy
PINWHEEL GALAXY (M101)	160,000 light-years	largest nearby galaxy
NGC 6872	750,000 light-years	largest known galaxy
COMA CLUSTER	3,000,000 light-years	largest nearby galaxy cluster
VIRGO SUPERCLUSTER	150,000,000 light-years	our neighbourhood galaxy supercluster
PERSEUS-PEGASUS SUPERCLUSTER	1 billion light-years	supercluster galaxy chain; largest known structure in the universe
UNIVERSE	30 billion light-years	

FURTHER READING

Magazines are the best resource for keeping up to date on astronomical discoveries. Most larger newsstands and libraries carry *Astronomy* and *Sky & Telescope*, the two major monthlies that more thoroughly cover the subject than any other single source. For more comprehensive background, these books are among my favourites.

General Astronomy

Journey Through the Universe by Jay M. Pasachoff (Saunders, New York, 1992). Beautifully illustrated introductory college text. Very up to date.

Astronomy: The Cosmic Journey by William K. Hartmann (Wadsworth, Belmont, California, many editions). Outstanding college text that makes a fine reference book.

Voyage Through the Universe by Time-Life Books. A stunning 20-volume series (nearly 3,000 pages), produced between 1988 and 1991, covering all aspects of astronomy. Lavishly illustrated. Available only by subscription to the series. For information, write Time-Life Books, Box C-32068, Richmond, VA 23261-2068.

The Astronomical Companion by Guy Ottewell (Astronomical Workshop, Greenville, South Carolina, 1979). Imaginative guide to understanding where we are in the galaxy and the universe. Not in bookstores; order from the publisher: Astronomical Workshop, Dept. of Physics, Furman University, Greenville, SC 29613.

Planets

The New Solar System edited by J. Kelly Beatty and Andrew Chaikin (Cambridge University Press, New York, 1990). The leading experts on planets, asteroids, comets, and so on, expound on their specialties. Well illustrated and nicely reproduced.

The Planetary System by David Morrison and Tobias Owen (Addison-Wesley, Reading, Massachusetts, 1988). Comprehensive reference.

Stars and Galaxies

Cycles of Fire by William K. Hartmann and Ron Miller (Workman, New York, 1987). A wonderfully illustrated book about the evolution of stars and galaxies.

Colliding Galaxies by Barry Parker (Plenum Press, New York, 1990). A look at modern research on galaxies.

Supernova! by Donald Goldsmith (St. Martin's Press, New York, 1989). What was learned from the 1987 supernova in the Large Magellanic Cloud.

Cosmology

Cosmology: The Science of the Universe by Edward R. Harrison (Cambridge University Press, New York, 1981).

Excellent background material on relativity, redshifts and the history of cosmology. Authoritative and thorough.

Coming of Age in the Milky Way by Timothy Ferris (Morrow, New York, 1988). Engaging and sometimes profound history of cosmology. Highly recommended.

The Big Bang by Joseph Silk (W.H. Freeman, New York, 1989). A readable, nonmathematical cosmology textbook.

The Dark Matter by Wallace and Karen Tucker (Morrow, New York, 1988). Easily digested introduction to the dark-matter enigma.

The Fifth Essence by Lawrence Krauss (Basic Books, New York, 1989). Well-written explanation of the search for dark matter.

Cosmologists

These books explore cosmology by focusing on the scientists who do—or did—the work. Highly recommended.

The Red Limit by Timothy Ferris (Morrow, New York, revised edition, 1983). The first and in many ways still the best of its type.

Origins: The Lives and Worlds of Modern Cosmologists by Alan Lightman and Roberta Brawer (Harvard University Press, Cambridge, Massachusetts, 1990).

Lonely Hearts of the Cosmos by Dennis Overbye (Harper Collins, New York, 1991). Superb.

Extraterrestrial Life

The Search for Extraterrestrial Intelligence by Thomas McDonough (Wiley, New York, 1987). Presents all the standard arguments and background material in a very readable fashion.

First Contact edited by Ben Bova and Byron Preiss (Plume/Penguin, New York, 1990). A collection of essays on the search for extraterrestrials by Isaac Asimov, Arthur C. Clarke, Philip Morrison and others.

Other Resources

For catalogues of slides, videos, posters, et cetera, contact any of the following: The Planetary Society, 110 S. Euclid Avenue, Pasadena, CA 91101; Astronomical Society of the Pacific, 390 Ashton Avenue, San Francisco, CA 94112; Hansen Planetarium Publications, 1098 S. 200 West, Salt Lake City, UT 84101; Sky Publishing, Box 9111, Belmont, MA 02178.

FRONTISPIECE: National Optical Astronomy Observatories, Kitt Peak National Observatory; p. 5 Anglo-Australian Telescope Board © 1981; p. 6 Canada-France-Hawaii Telescope Corp.; p. 8 Jet Propulsion Laboratory.

CHAPTER 1: p. 10 Royal Observatory, Edinburgh © 1980; p. 12-13 Illustration by John Bianchi; p. 14 NASA; p. 15 NASA; p. 16 (top) National Optical Astronomy Observatories, Kitt Peak National Observatory; p. 16 (bottom) Leo Henzl; p. 17 Illustration by Victor Costanzo © 1986; p. 18 Anglo-Australian Telescope Board © 1978.

CHAPTER 2: p. 20 Illustration by Adolf Schaller; p. 22 (left) NASA; p. 22 (right) Illustration by Adolf Schaller; p. 23 Illustration by Adolf Schaller; p. 24 Illustration by Don Davis; p. 25 Lunar and Planetary Institute; p. 26 (top, bottom left and right) Lunar and Planetary Institute; p. 27 Illustration by David Egge; p. 29 U.S. Geological Survey/NASA; p. 30 U.S. Geological Survey/NASA; p. 31 Ames Research Center, NASA; p. 32 Jet Propulsion Laboratory; p. 33 Jet Propulsion Laboratory; p. 34 (top, bottom) Illustrations © James Hervat; p. 35 Jet Propulsion Laboratory; p. 36 Jet Propulsion Laboratory; p. 37 Illustration by Don Davis.

CHAPTER 3: p. 38 Illustration by Adolf Schaller; p. 40-41 (all) Space Telescope Science Institute; p. 42 Jet Propulsion Laboratory; p. 43 Illustration by Adolf Schaller © Edmund Scientific; p. 44 Illustration by Adolf Schaller; p. 45 (top) Illustration by Adolf Schaller © Edmund Scientific; p. 45 (centre, bottom) Jet Propulsion Laboratory; p. 46 Space Telescope Science Institute; p. 47 Illustration by David Egge; p. 48 (top, bottom) Jet Propulsion Laboratory; p. 49 Illustration by Don Davis; p. 50 Jet Propulsion Laboratory; p. 51 Jet Propulsion Laboratory.

CHAPTER 4: p. 52 Illustration by Adolf Schaller; p. 54 (top, centre, bottom) Jet Propulsion Laboratory; p. 55 (left) Illustration by Michael Carroll; p. 55 (right) Illustration by David Egge; p. 56 Jet Propulsion Laboratory; p. 57 (left) Illustration by Michael Carroll; p. 57 (right) Jet Propulsion Laboratory; p. 58 Illustration by Adolf Schaller; p. 59 (left) Illustration by Michael Carroll; p. 59 (right) Jet Propulsion Laboratory; p. 60-61 (all) Jet Propulsion Laboratory; p. 62 Illustration © James Hervat; p. 63 (left) Illustration by Michael Carroll; p. 63 (right) European Space Agency; p. 64 Jet Propulsion Laboratory; p. 65 Jim Riffle, Astro-Works.

CHAPTER 5: p. 66 Royal Observatory, Edinburgh © 1980; p. 68 (left) John Hicks; p. 68 (right) Illustration by Adolf Schaller; p. 69 (left) Illustration by Michael Carroll; p. 69 (right) NASA; p. 70 Illustration by David Egge; p. 71 Illustration by Victor Costanzo © Edmund Scientific; p. 72-73 Space Telescope Science Institute; p. 74 Anglo-Australian Telescope Board © 1979; p. 77 Illustration by John Bianchi; p. 78 Illustration by Adolf Schaller © Edmund Scientific; p. 79 Laird Thompson, Canada-France-Hawaii Telescope; p. 80 Karl Schwarzschild Observatory; p. 82 Illustration by Adolf Schaller; p. 83 University of Toronto; p. 84 Illustration by Adolf Schaller; p. 85 Illustration by Adolf Schaller.

CHAPTER 6: p. 86 Michael Watson; p. 88 Illustration by Brian Sullivan; p. 89 (left) Mount Wilson and Las Campanas Observatories, Carnegie Institute of Washington; p. 89 (right) Jet Propulsion Laboratory/California Institute of Technology; p. 90 (bottom) Michael Watson; p. 90-91 Illustration by Don Davis and Geoffery Chandler © 1986; p. 91 (bottom left) Anglo-Australian Telescope Board © 1980; p. 91 (bottom right) Smithsonian Astrophysical Observatory; p. 92 Royal Observatory, Edinburgh © 1984; p. 93 (top, bottom) Anglo-Australian Telescope Board © 1980; p. 94 (left) Hale Observatories; p. 94 (right) National Optical Astronomy Observatories, Kitt Peak National Observatory; p. 95 Smithsonian Astrophysical Observatory; p. 96 (top) Terence Dickinson; p. 96 (bottom) © Fred Espenak; p. 97 (clockwise from top left) Laird Thompson, Canada-France-Hawaii Telescope; © Fred Espenak; Smithsonian Astrophysical Observatory; Smithsonian Astrophysical Observatory; Jet Propulsion Laboratory; Jet Propulsion Laboratory; Smithsonian Astrophysical Observatory; Lick Observatory; p. 97 (right) Smithsonian Astrophysical Observatory; p. 99 Illustration by Adolf Schaller; p. 100 National Optical Astronomy Observatories, Kitt Peak National Observatory; p. 101 Illustration by David Egge; p. 102 (left) Space Telescope Science Institute; p. 102 (right) Antony Tyson, Bell Laboratories; p. 103 Space Telescope Science Institute.

CHAPTER 7: p. 104 Anglo-Australian Telescope Board © 1977; p. 106 (top) M. Seldner, B. Siebers, Edward Groth, P.J.E. Peebles; p. 106 (bottom) National Optical Astronomy Observatories, Kitt Peak National Observatory; p. 107 Lowell Observatory; p. 108 Illustration by Victor Costanzo; p. 109 Smithsonian Astrophysical Observatory; p. 110-11 Illustration by Adolf Schaller; p. 112 (left) National Optical Astronomy Observatories, Cerro Tololo Inter-American Observatory; p. 112 (right) R.B. Tully; p. 113 Illustration by Adolf Schaller; p. 115 Illustration by John Bianchi; p. 116-17 Alan Carruthers; p. 118 Terence Dickinson; p. 119 (top right) Terence Dickinson; p. 119 (bottom right and left, top left) Alan Carruthers.

CHAPTER 8: p. 120 Illustration by Adolf Schaller; p. 122-23 Karl Schwarzschild Observatory; p. 124 (left, right) Illustrations by Adolf Schaller © Edmund Scientific; p. 125 Illustration by Don Davis © 1986; p. 126 Alan Dyer © 1982; p. 127 (left) Illustration by David Egge; p. 127 (right) Illustration by Adolf Schaller © Edmund Scientific; p. 128 Jet Propulsion Laboratory; p. 129 Illustration by David Egge; p. 131 Illustration by Don Davis.

EPILOGUE: p. 132 Anglo-Australian Telescope Board © 1981; p. 135 J.R. Angel, University of Arizona, Steward Observatory; p. 137 NASA; p. 139 Adolf Schaller; p. 140-41 Adolf Schaller; p. 156 Anthea Weese.

INDEX

THE AUTHOR

Terence Dickinson is Canada's leading astronomy writer. He is the author of nine books, including the best-selling NightWatch, *and has written hundreds of magazine articles for publications ranging from* Reader's Digest *to* Sky & Telescope. *He also writes a weekly astronomy column for* The Toronto Star, *teaches astronomy part-time at St. Lawrence College, Kingston, Ontario, and is an astronomy commentator for "Quirks and Quarks," CBC Radio's weekly science programme. Before turning to science writing full-time in 1976, he was editor of* Astronomy *magazine and was an instructor at several science museums and planetariums in Canada and the United States. In 1992, he received the Royal Canadian Institute's Sandford Fleming Medal for achievements in advancing public understanding of science. He and his wife Susan live near the village of Yarker in rural eastern Ontario.*